CATASTROPHIC DISASTERS IN AMERICA: PREPAREDNESS, RESPONSE, AND RECOVERY SYSTEM

Catastrophic Disasters in America: Preparedness, Response, and Recovery System

General Accountability Office

Nova Science Publishers, Inc.
New York

For permission to use material from this book please contact us:
Telephone 631-231-7269; Fax 631-231-8175
Web Site: http://www.novapublishers.com

Library of Congress Cataloging-in-Publication Data

Catastrophic disasters in America : preparedness, response, and recovery system / General Accountability Office (editor).

p. cm.

ISBN 978-1-60456-070-0 (hardcover)

1.Emergency management—United States. 2. Preparedness—Government policy—United States. 3. Disaster relief—United States—Finance. I. United States. Government Accountability Office.

HV551.3.C38 2008
502.C466 2008
363.340973--dc22

2007039573

Published by Nova Science Publishers, Inc. ✣ New York

CONTENTS

PREFACE

The three basic elements in preparing for, responding to and recovering from any catastrophic disaster are (1) leadership; (2) capabilities; and (3) accountability. Leadership in the form of legal authorities, roles and responsibilities, and lines of authority at all levels of government must be clearly defined, effectively communicated, and well understood in order to facilitate rapid and effective decision making.

Developing the capabilities needed for catastrophic disasters should be part of an overall national effort designed to integrate and define what needs to be done, where, by whom, and how well. Ensuring needed capabilities are ready requires effective planning and coordination, plus robust training and exercises in which the capabilities are realistically tested, problems identified, and subsequently addressed in partnership with federal, state, local, and nongovernmental stakeholders. In addition, integrating an allhazards risk management framework into decision making is central to assessing catastrophic disaster risks and guiding the development of national capabilities to prevent or mitigate where possible and respond to such risks.

This new book examines crucial issues in dealing with catastrophic disasters in America.

ABBREVIATIONS

DCMS	Disaster Credit Management System
DOD	Department of Defense
DHS	Department of Homeland Security
ESF	Emergency Support Function
FCO	Federal Coordinating Officer
FEMA	Federal Emergency Management Agency
HHS	Health and Human Services
HSPD	Homeland Security Presidential Directive
IHP	Individuals and Households Program
JFO	Joint Field Office
NDMS	National Disaster Medical System
NFC	National Finance Center
NFIP	National Flood Insurance Program
NIMS	National Incident Management System
NOAA	National Oceanographic and Atmospheric Administration
NRP	National Response Plan
PFO	Principal Federal Officer
SBA	Small Business Administration

Chapter 1

INTRODUCTION*

WHY GAO DID THIS STUDY

Hurricane Katrina was the largest, most destructive natural disaster in our nation's history. The problems experienced in responding to Katrina resulted in a number of investigations—by congressional committees, the White House Homeland Security Council, and others--regarding the preparations for and response to Katrina. GAO assisted the congressional investigations and, under the Comptroller General's authority, initiated a number of Katrina- related reviews. In March 2006 testimony, GAO provided its preliminary observations to Congress.

The purpose of this report is to summarize what went well and why, what did not go well and why, and what changes are needed to improve the nation's readiness to respond to a catastrophic disaster; and to identify selected issues associated with the Gulf Coast's recovery. This report is based on GAO's prior work on catastrophic disasters, including Hurricane Andrew in 1992, the over 30 GAO reports completed to date on Hurricanes Katrina and Rita, ongoing GAO work, and other Hurricane Katrina reviews and lessons learned.

WHAT GAO RECOMMENDS

This report includes six recommendations to the Secretary of Homeland Security with which DHS generally agreed, describing actions taken to implement them. The report also includes a matter for congressional consideration.

WHAT GAO FOUND

Hurricane Katrina was a catastrophic disaster whose scope and destruction severely tested all levels of governments in the affected areas and the nation as a whole. It almost immediately overwhelmed state and local first responders, and the response required outside action and support from many sources. The heroic efforts by many saved thousands of lives.

The federal government, many states, local governments, plus nonprofit and private sector organizations provided substantial personnel and resources to assist in the response, but these proved insufficient to meet the immediate challenges posed by Hurricane Katrina's effects.

The three basic elements in preparing for, responding to and recovering from any catastrophic disaster are (1) leadership; (2) capabilities; and (3) accountability. Leadership in the form of legal authorities, roles and responsibilities, and lines of authority at all levels of government must be clearly defined, effectively communicated, and well understood in order to facilitate rapid and effective decision making. DHS has made revisions to the National Response Plan designed to further clarify federal roles and responsibilities, but their effect has not yet been tested in an actual disaster.

Developing the capabilities needed for catastrophic disasters should be part of an overall national effort designed to integrate and define what needs to be done, where, by whom, and how well. Ensuring needed capabilities are ready requires effective planning and coordination, plus robust training and exercises in which the capabilities are realistically tested, problems identified, and subsequently addressed in partnership with federal, state, local, and nongovernmental stakeholders. In addition, integrating an all-hazards risk management framework into decision making is central to assessing catastrophic disaster risks and guiding the development of national capabilities to prevent or mitigate where possible and respond to such risks. DHS has announced a number of actions to improve readiness and response for catastrophic disasters, but there is little information available on the extent to which these changes are operational.

Accountability controls and mechanisms ensure that resources are used appropriately for valid purposes. Following a catastrophic disaster, decision-makers face a tension between the demand for rapid response and recovery assistance—including assistance to victims—and implementing appropriate controls and accountability mechanisms. Our work and that of others found, for example, the processes for confirming disaster victims' eligibility for assistance were insufficient and resulted in millions of dollars in questionable payments to fraudulent claimants. Also, some contracts had insufficient provisions to ensure that prices were fair and reasonable. DHS has reported that it has taken steps to address some of the concerns, including working to complete more contracts for key services in advance of a disaster and improving its ability to verify individual claimant eligibility for disaster benefits and assistance.

Just over 1 year ago, Hurricane Katrina made final landfall in coastal Louisiana and Mississippi, and its destructive force extended to the western Alabama coast. Hurricane Katrina was the worst natural disaster in our nation's history in geographic scope, the extent and severity of its destruction and damage, and the number of persons displaced from their homes. Hurricane Katrina and the following Hurricanes Rita and Wilma—also among the most powerful hurricanes in the nation's history—graphically illustrated the limitations at that time of the nation's readiness and ability to respond effectively to a catastrophic disaster—that is, a disaster whose effects almost immediately overwhelm the response capacities of affected state and local first responders and require outside action and support from the federal government and other entities. Reports by the House Select Bipartisan Committee to Investigate the Preparation for and Response to Hurricane Katrina,[1] the Senate Homeland Security and Governmental Affairs Committee,[2] the White House Homeland Security

* Excerpted from GAO Report 06-618, dated September 2006.

Council,[3] the Department of Homeland Security (DHS) Inspector General,[4] and DHS and the Federal Emergency Management Agency (FEMA)[5] have all identified a variety of failures and some strengths in the preparations for, response to, and initial recovery from Hurricane Katrina. Collectively, these reports, along with GAO's various reports and testimonies, have chronicled the strengths and weaknesses of federal, state, local, and nongovernmental efforts, and offered a number of specific recommendations for improving the nation's ability to effectively prepare for and respond to catastrophic disasters. Legislation to reorganize the federal government's emergency response responsibilities has been introduced in both the House and the Senate and the administration has undertaken several efforts to address disaster management weaknesses identified in the aftermath of the 2005 hurricane season.

The 2005 hurricane season was particularly active, and the National Oceanographic and Atmospheric Administration (NOAA) estimated that there was a 75-percent probability of another active hurricane season in 2006 (the hurricane season runs from June 1st through November 30th each year). However, NOAA reminds us that hurricane-spawned catastrophic disasters can occur even in years with near-normal or below-normal levels of activity. The nation's second most destructive hurricane—Andrew in 1992—occurred during a season with otherwise below normal activity. Moreover, the nation must be prepared to respond to and recover from a wide range of other disasters that may occur at any time of year and could be the result of nature or human action, such as a possible influenza pandemic or terrorist attack.

In a March 2006 testimony,[6] we provided our preliminary observations regarding the preparedness for, response to, and recovery from Hurricanes Katrina and Rita. We noted that the key issues were in many ways very reminiscent of the issues identified in the wake of Hurricane Andrew in 1992[7] and fell into three broad, interrelated categories: (1) leadership, including clearly defined roles and responsibilities of all key participants; (2) building and sustaining effective capabilities through coordinated planning, training, and exercises; and (3) maintaining accountability for the use of resources while providing assistance and resources as quickly as possible. The overall objective of this work was (1) to provide Congress with a summary of what went well and why, what did not go well and why, and what specific changes are needed to improve this nation's emergency preparedness, response, and recovery system; and (2) identify selected issues associated with the Gulf Coast's recovery. Following the themes from our prior work, this report discusses the major findings from our recent and ongoing work on the aftermath of Hurricane Katrina and relevant findings from Congress and the administration with respect to (1) the actions of government agencies during Hurricane Katrina that made positive contributions to the response and recovery and those that were less positive; (2) catastrophic disaster leadership roles and responsibilities; (3) capabilities to prepare for, respond to, and recover from catastrophic disasters; (4) recognizing the tension between the need for timely action and the need for appropriate controls and accountability mechanisms; and (5) selected longer-term recovery issues, including the rebuilding effort along the Gulf Coast.

This report is based on our work on Hurricanes Katrina and Rita and our visits to the areas most affected by Hurricanes Katrina and Rita—Alabama, Louisiana, Mississippi, and Texas. We interviewed various officials, including senior federal officials, the governors of Alabama, Louisiana, Mississippi, and Texas, and the mayor of New Orleans. We also analyzed information from the various involved federal agencies, such as FEMA and the Department of Defense (DOD) and the federal audit community; state organizations including state emergency management agencies; National Guard officials in the states, state agencies and state auditors; local officials; and representatives from nongovernmental agencies. We also have studied the findings in reports issued by Congress, DHS/FEMA, DHS's Office of Inspector General, and the White House Homeland Security Council.

In addition, we have an extensive body of work on prior catastrophic disasters. For example, we drew upon several past reviews in 1993, examining the federal response to Hurricane Andrew. We also conducted extensive work following the events of September 11, 2001,[8] and on tsunami preparedness and recovery issues.[9] Our reports focused on improving the immediate response to catastrophic disasters that raise unique challenges, and we made various recommendations within this context, many of which continue to apply and help form the basis of our views today, including the issue of FEMA's future organizational placement. In the months following Hurricane Katrina, we provided support to congressional investigations of the preparedness for and response to Hurricane Katrina and have issued reports on such topics as the military's role in catastrophic disasters;[10] contracting practices for response and recovery activities;[11] the evacuation of hospitals, nursing homes and other vulnerable populations;[12] charitable assistance and the coordination between FEMA and the Red Cross;[13] ensuring the appropriate use and accountability for international assistance;[14] and issues associated with guarding against fraud in the provision of expedited assistance to disaster victims.[15] To date we have published over 30 GAO reports and testimonies on Hurricane Katrina-related matters. We will continue to focus on ways to improve the nation's readiness and capacity to respond to catastrophic disasters as well as issues associated with the long-term recovery of the Gulf Coast region, including housing, health care, levee reconstruction, and economic recovery.

We conducted our work from September 2005 through August 2006, in accordance with generally accepted government auditing standards.

RESULTS IN BRIEF

While much of the publicity regarding Hurricane Katrina has focused on what did not go well in connection with the governments' activities, there are notable exceptions. For example, the National Weather Service and the National Hurricane Center provided accurate forecasts of the time and location of Hurricane Katrina's landfall, accompanied by warnings of the hurricane's potential destructive force. Dedicated and heroic efforts by many, including local first responders, hospital personnel, the U.S. Coast Guard, the National Guard, active duty military troops, and volunteers saved thousands of lives. Federal agencies such as the Postal Service, the Social Security Administration, the National Finance Center, the Forest Service, and the Internal Revenue Service demonstrated their flexibility and adaptability, responding to Katrina's challenges. Many states sent supplies, first responders, National Guard personnel, and other resources to assist the areas hardest hit by the disasters. Many charitable, faith-based and private sector organizations supplemented governmental efforts providing food and shelter for thousands. Despite these efforts, various reports and our own work on the 2005 catastrophic disasters suggest that the substantial human, financial, and technological resources and capabilities marshaled by governments at all levels were inadequate to meet the immediate challenges posed by the disaster's unprecedented geographic scope, degree of damage, and the resulting number of hurricane victims who had to be relocated.

In preparing for, responding to, and recovering from any catastrophic disaster, the legal authorities, roles and responsibilities, and lines of authority at all levels of government must be clearly defined, effectively communicated, and well understood in order to facilitate rapid and effective decision making. The experience of Hurricane Katrina showed the need to improve leadership at all levels of government in order to better respond to a catastrophic disaster. For example, there were problems experienced with roles and responsibilities under the National Response Plan (NRP),[16] and ambiguities concerning what constituted an incident of national significance to trigger the NRP as well as what constituted a catastrophic incident to trigger the proactive response of the NRP's Catastrophic Incident Annex. On May 25, 2006, DHS released changes to the NRP regarding leadership issues, such as which situations require secretarial leadership; the process for declaring incidents of national significance; and the scope of the NRP and its Catastrophic Incident Annex. The revised NRP clearly states that the Secretary of Homeland Security, who reports directly to the President, is responsible for declaring and managing incidents of national significance, including catastrophic incidents. DHS has not yet published the final supplement to the Catastrophic Incident Annex, which is intended to provide more detail for implementing the annex. Until the supplement is final and distributed, agencies that have responsibilities under the supplement cannot complete the implementation plans and agreements needed to make the annex and supplement fully operational. The White House Homeland Security Council report included 44 recommendations that were intended to be implemented quickly, of which 18 are focused on improving and clarifying the legal authorities, roles and responsibilities, and lines of authority. DHS has provided limited information on the status of its implementation of the White House recommendations, although it has reported actions taken on some issues raised in the White House Homeland Security Council report and in other critiques. For example, DHS has designated principal federal officials and federal coordinating officers for regions and states at-risk of hurricanes and further described their respective roles in coordinating disaster response—a source of some confusion in the federal response to Hurricane Katrina. However, the changes may not have fully resolved the leadership issues with regard to the roles of the principal federal officer and federal coordinating officer. While the Secretary of Homeland Security may avoid conflicts by appointing a single individual to serve in both positions in non-terrorist incidents, confusion may persist if the Secretary of Homeland Security does not exercise this discretion to do so. Furthermore, this discretion does not exist for terrorist incidents, and the revised NRP does not specifically provide a rationale for this limitation. Congress also has raised concerns that FEMA's performance problems exhibited during the hurricane response may stem from its organizational placement and its budgetary relationship within DHS. As we stated in our March 8 and May 9, 2006, testimonies, organizational changes alone, while potentially important, will not be enough to adequately address the underlying systemic conditions that resulted in FEMA's performance problems during the last hurricane season. In our view, a number of other factors may be ultimately more important to FEMA's success in responding to and recovering from future disasters, including catastrophic disasters, than its organizational placement. We have identified several factors that Congress should consider as it deliberates FEMA's organizational placement: (1) mission relevancy, (2) similarity of goals and objectives, (3) the ability to leverage the effectiveness of other agencies, and (4) gains through consolidation.

As stated earlier, numerous reports and our own work suggest that the substantial resources and capabilities marshaled by state, local, and federal governments and nongovernmental organizations were insufficient to meet the immediate challenges posed by the unprecedented degree of damage and the resulting number of hurricane victims caused by Hurricanes Katrina and Rita. Developing the capabilities needed for catastrophic disasters should be part of an overall national preparedness effort that is designed to integrate and define what needs to be done, where, based on what standards, how it should be done, and how well it should be done. The nation's experience with Hurricanes Katrina and Rita reinforces some of the questions surrounding the adequacy of capabilities in the context of a catastrophic disaster—particularly in the areas of (1) situational assessment and awareness; (2) emergency communications; (3) evacuations; (4) search and rescue; (5) logistics; and (6) mass care and sheltering. Ensuring that needed capabilities such as these are available requires effective planning and coordination, as well as training and exercises in which the capabilities are realistically tested and problems identified and subsequently addressed in partnership with other federal, state, and local stakeholders, as capabilities are built upon the appropriate combination of people, skills, processes, and assets. In addition, ongoing work is still needed by DHS to address FEMA's human resource challenges. Finally, as we stated in our March 2006 testimony, the use of a risk management methodology—integrating systematic concern for risk into the normal cycle of agency decision making and implementation—should be central to assessing the risk for catastrophic disasters, guiding the development of national capabilities and the expertise that can be used to respond effectively to catastrophic disasters.

Although controls and accountability mechanisms help to ensure that resources are used appropriately, during a catastrophic disaster decision-makers struggle with the tension between implementing controls and accountability mechanisms and the demand for rapid response and recovery assistance. On one hand, our work found many examples where quick action could not occur due to procedures that required extensive, time-consuming processes, delaying the delivery of vital supplies and other assistance. On the other hand, we also found examples where FEMA's processes under assistance programs to disaster victims left the federal government vulnerable to fraud and the abuse of expedited assistance payments. We estimate that through February 2006, FEMA made about $600 million to $1.4 billion in improper and potentially fraudulent payments to applicants who used invalid information to apply for expedited cash assistance. DHS and FEMA have reported a number of actions that are to be in effect for the hurricane season so that federal recovery programs will have more capacity to rapidly handle a catastrophic incident but also provide accountability. Examples include significantly increasing the quantity of pre-positioned supplies, such as food, ice, and water; placing global positioning systems on supply trucks to track their location and better manage the delivery of supplies; an enhanced phone system for victim assistance applications that can handle up to 200,000 calls per day; and improved computer systems and processes for verifying the eligibility of those applying for assistance. Effective implementation of these and other planned improvements will be critical to achieving their intended outcomes. Finally, catastrophic disasters not only require a different magnitude of capabilities and resources for effective response, they may also require more flexible policies and operating procedures. In a catastrophe, streamlining, simplifying, and expediting decision making should quickly replace "business as usual" and the unquestioned following of long-standing policies and operating procedures used in normal situations for providing relief to disaster victims. At the same time, controls and accountability mechanisms must be sufficient to provide the documentation needed for expense reimbursement and reasonable assurance that resources have been used legally and for the purposes intended.

The federal government also will be a major partner in the longer-term recovery and rebuilding of individuals and communities along the Gulf Coast. Among the areas requiring federal attention are (1) assessing the environmental hazards created by the storms; (2) rebuilding and strengthening the levees; (3) providing assistance to school districts that have enrolled large numbers of evacuee children; and (4) building the capacity to address demand in multiple victims assistance programs such as financial assistance or loans for repair and replacement of housing and the rebuilding of businesses.

This report summarizes recommendations we have made in our work, some following Hurricane Andrew in 1992 and others in the aftermath of Hurricanes Katrina and Rita. We are also updating and formalizing several recommendations initially presented in our March 8, 2006, testimony. Specifically, we are recommending that DHS (1) rigorously re-test, train, and exercise its recent clarification of the roles, responsibilities, and lines of authority for all levels of leadership, implementing changes needed to remedy identified coordination problems; (2) direct that the NRP base plan and its supporting Catastrophic Incident Annex be supported by more robust and detailed operational implementation plans; (3) provide guidance and direction for federal, state, and local planning, training, and exercises to ensure such activities fully support preparedness, response, and recovery responsibilities at a jurisdictional and regional basis; (4) take a lead in monitoring federal agencies' efforts to prepare to meet their responsibilities under the NRP and the interim National Preparedness Goal; (5) use a risk management approach in deciding whether and how to invest finite resources in specific capabilities for a catastrophic disaster; and (6) provide guidance on advanced procurement practices and procedures for those federal agencies with roles and responsibilities under the NRP so that these agencies can better manage disaster-related procurement, and establish an assessment process to monitor agencies' continuous planning efforts for their disaster-related procurement needs and the maintenance of capabilities. We are also recommending that Congress give federal agencies explicit authority to take actions to prepare for all types of catastrophic disasters when there is warning. We also offer some factors that Congress may wish to consider as it carries out its oversight and legislative responsibilities with regard to national preparedness and the recovery of the Gulf Coast region.

Chapter 2

Background

Hurricanes Katrina and Rita left more than 1,500 dead, affected over 90,000 square miles, caused more than $80 billion in damage, and forced mass evacuations from five states along the Gulf Coast, according to DHS. An estimated 600,000 households were displaced from affected areas and 50,000 to 100,000 households remained in temporary housing 6 months later. As a result, 44 states and the District of Columbia received hundreds of thousands of evacuees, providing them with care and shelter over an extended period. These events tested the nation's ability to respond to catastrophic events and demonstrated the importance of ensuring the effectiveness of federal, state, and local emergency response plans and the ability to quickly synchronize intergovernmental efforts. This catastrophic disaster provided a sobering picture of the overwhelming strains on disaster response and recovery capacities if there are back-to-back catastrophic disasters in the same area.

Significant local, state, and federal resources were mobilized to respond to the Hurricane Katrina disaster, along with significant participation from charitable and private-sector organizations. Among federal agencies, the National Weather Service and its National Hurricane Center were proactive and vigilant, accurately predicted and tracked the size, scale, and path of Hurricane Katrina, and regularly contacted local, state and federal leaders to apprise them of the situation. The Coast Guard, National Guard, and state and local responders acted quickly to implement search and rescue efforts that saved thousands of stranded and desperate victims. The National Guard and the active military also played a major role in saving lives, providing food and shelter, and transporting victims who needed immediate medical care. Federal agencies such as the Postal Service, Social Security Administration, National Finance Center, Forest Service, and Internal Revenue Service demonstrated their flexibility in performing needed activities and services, and responding in the face of many challenges. However, the capabilities of several federal, state, and local agencies were clearly overwhelmed. As events unfolded in the immediate aftermath and ensuing days after Hurricane Katrina's final landfall, responders at all levels of government—many victims themselves—encountered significant breakdowns in vital areas such as emergency communications as well as obtaining and deploying essential supplies and equipment.

Stafford Act is the Principal Federal Statute on Federal Disaster Assistance to States and Localities

There are several federal legislative and executive provisions that support preparation for and response to emergency situations. The Robert T. Stafford Disaster Relief and Emergency Assistance Act (the Stafford Act)[17] primarily establishes the programs and processes for the federal government to provide major disaster and emergency assistance to states, local governments, tribal nations, individuals, and qualified private nonprofit organizations. Upon a governor's request, the President can declare an "emergency" or a "major disaster" under the Stafford Act, which triggers specific types of federal relief. The Stafford Act defines an emergency as any occasion or instance for which, in the determination of the President, federal assistance is needed to supplement state and local efforts and capabilities to save lives and to protect property and public health and safety, or to lessen or avert the threat of a catastrophe in any part of the United States. Under an emergency declaration, the federal government has authority to engage in various emergency response activities, such as debris removal, temporary housing assistance, and the distribution of medicine, food, and other consumables. The Stafford Act places a $5 million limit on federal emergency assistance, but the President may exceed the limit, followed by a report to Congress.

The Stafford Act defines a "major disaster" as any natural catastrophe or, regardless of cause, any fire, flood, or explosion, in any part of the United States, which the President determines causes damage of sufficient severity and magnitude to warrant major disaster assistance under the Stafford Act to supplement the efforts and available resources of states, local governments, and disaster relief organizations in alleviating damage, loss, hardship, or suffering. Under a major disaster declaration, the federal government has the authority to engage in the same activities authorized under an emergency declaration, but without the $5 million ceiling. In addition, major disaster assistance includes a variety of assistance not available in the context of an emergency declaration. For example, in a major disaster, the federal government may provide unemployment assistance, food coupons to low-income households, and repair, restoration and replacement of certain damaged facilities, among other things.

For Hurricane Katrina, the President issued emergency declarations under the Stafford Act for Louisiana on August 27, 2005, and Mississippi and Alabama on August 28, 2005. The President made major disaster declarations for Florida on August 28, 2005, and Louisiana, Mississippi, and Alabama on August 29, 2005, the same day that Hurricane Katrina made final landfall in Plaquemines Parish, Louisiana at 6:10 a.m. and about 4 hours later at the Louisiana/Mississippi border.

Federal Disaster Assistance and Relief is to Be Administered by FEMA under the Stafford Act

FEMA, within DHS, has responsibility for administering the provisions of the Stafford Act, the principal federal statute governing federal disaster assistance and relief. FEMA was an independent agency from 1979 until 2003 and was a cabinet-level agency from 1996 to 2003. FEMA became part of the newly formed DHS in March 2003. Prior to FEMA's creation in 1979, federal disaster response and recovery was managed by an agency within the Department of Housing and Urban Development (HUD). The 1960s and early 1970s brought massive disasters requiring major federal response and recovery operations by the Federal Disaster Assistance Administration, the department within HUD responsible at that time for disaster response and recovery. Hurricane Carla struck in 1962, Hurricane Betsy in 1965, Hurricane Camille in 1969, and Hurricane Agnes in 1972. A major earthquake hit Alaska in 1964, and the San Fernando earthquake rocked Southern California in 1971. To respond to national concern regarding these events, Congress passed the 1974 Disaster Relief Act that established the process of presidential disaster declarations. However, emergency and disaster activities were still fragmented. Many parallel programs and policies existed at the state and local level, compounding the complexity of federal disaster relief efforts. In 1979, President Carter issued an executive order that merged many of the separate disaster-related responsibilities into a new, independent Federal Emergency Management Agency. Among other agencies, FEMA absorbed the Federal Insurance Administration, the National Fire Prevention and Control Administration, the National Weather Service Community Preparedness Program, the Federal Preparedness Agency of the General Services Administration, and the Federal Disaster Assistance Administration activities from HUD. Civil defense responsibilities were also transferred to the new agency from the Defense Department's Defense Civil Preparedness Agency.

Interstate Disaster Assistance Was Also Provided through the Emergency Management Assistance Compact

In addition to resources provided by the federal government, states affected by a catastrophic disaster can also turn to other states for assistance in obtaining needed surge capacity—the ability to draw on additional resources, such as personnel and equipment, needed to respond to and recover from the incident. One way of sharing personnel and equipment across state lines is through the use of an interstate compact that provides a legal and administrative framework for such emergency requests. Following the devastation of Hurricane Andrew in 1992, the Southern Governors' Association, along with the Virginia Department of Emergency Services, formed the Southern Regional Emergency Management Assistance Compact (SREMAC) in August 1993. In 1995, SREMAC opened membership to any state or territory, becoming the Emergency Management Assistance Compact (EMAC), approved by Congress in 1996.[18] As of September 2005, EMAC included all 50 states, the District of Columbia, Puerto Rico, and the U.S. Virgin Islands and has been activated during a variety of emergency events including hurricanes, floods, wildfires, and the September 11 terrorist attacks. Administered by the National Emergency Management Association (NEMA),[19] the compact establishes a structure for requesting and deploying assistance, reimbursing states that provide assistance, and conferring liability and workers' compensation protection. EMAC was not designed to supplant federal support nor does it obligate member states to provide resources to another state.

The deployment of resources through EMAC in response to Hurricanes Katrina, Rita, and Wilma was by far the largest use of state-to-state mutual assistance in U.S. history. According to EMAC, as of March 24, 2006, Louisiana and Mississippi had made almost 1,900 requests for assistance resulting in the deployment of 61,450 civilian and National Guard personnel.

The Homeland Security Act Requires a Comprehensive Approach to Disaster Preparation and Response

The Homeland Security Act of 2002[20] required the newly established DHS to develop a comprehensive National Incident Management System (NIMS). NIMS is intended to provide a consistent framework for incident management at all jurisdictional levels regardless of the cause, size, or complexity of the situation and to define the roles and responsibilities of federal, state, and local governments, and various first responder disciplines at each level during an emergency event. NIMS established the Incident Command System as a standard incident management organization with five functional areas—command, operations, planning, logistics, and finance and administration—for management of all major incidents. It also prescribes interoperable communications systems and preparedness before an incident happens, including planning, training, and exercises.

The Homeland Security Act of 2002 also required DHS to consolidate existing federal government emergency response plans into a single, integrated and coordinated national response plan. In December 2004, DHS issued the National Response Plan (NRP), intended to be an all-discipline, all-hazards plan establishing a single, comprehensive framework for the management of domestic incidents where federal involvement is necessary. The NRP does not apply to the majority of incidents occurring each year that are handled by local jurisdictions or agencies through established authorities and existing plans under the planning assumption that incidents arc typically managed at the lowest possible geographic, organizational, and jurisdictional level. Rather, the NRP is applicable to incidents that go beyond the state and local level and require a coordinated federal response. The NRP, operating within the framework of NIMS, provides the structure and mechanisms for national-level policy and operational direction for domestic incident management. The NRP also includes a Catastrophic Incident Annex, which describes an accelerated, proactive national response to catastrophic incidents. A draft of a more detailed and operationally specific Catastrophic Incident Supplement for the NRP's Catastrophic Incident Annex had not been approved at the time of Hurricane Katrina, although the NRP's 120-day schedule for implementing the supplement had passed. Once finalized, the supplement, as supported by agency-level implementation agreements and plans, is to serve as the operational framework for implementing the Catastrophic Incident Annex for all types of catastrophes.

DHS Policies and Plans Have Identified a Strategy for Building and Sustaining Preparedness, Response, and Recovery Capabilities

Capabilities—the ability to carry out specific tasks under particular conditions with desired results—are built upon the appropriate combination of people, skills, processes, and assets. Catastrophic disasters place particularly wide-ranging demands on emergency response capabilities. By their very nature, catastrophic disasters involve extraordinary levels of mass casualties, damage, or disruption that are likely to immediately overwhelm state and local responders. For non-catastrophic disasters, the federal government has historically been in a support and assist role, providing resources and other assistance to enable state and local governments to carry out their responsibilities. However, for catastrophic disasters that can overwhelm the ability of state and local and voluntary agencies to adequately provide victims with essential services, the federal government generally plays a more central role—providing selected resources where they are needed or likely to be needed. As we noted in our 1993 report following Hurricane Andrew, where there is warning of a catastrophic disaster, federal agencies need to mobilize resources and deploy personnel before the disaster strikes so that the federal government is ready and able to act quickly to assist the affected areas. However, current law—in particular the Stafford Act—does not explicitly authorize such pre-declaration activities.

Developing the capabilities needed for large-scale disasters is part of an overall national preparedness effort that should integrate and define what needs to be done, where, based on what standards, how it should be done, and how well it should be done. Along with the NRP and NIMS, DHS has developed the National Preparedness Goal, as required by Homeland Security Presidential Directive 8 (HSPD-8). Considered as a group, these three documents are intended to guide investments in emergency preparedness and response capabilities. The NRP describes what needs to be done in response to an emergency incident, the NIMS describes how to manage what needs to be done, and the National Preparedness Goal describes how well it should be done.

The interim National Preparedness Goal is particularly important for determining what capabilities are needed, especially for a catastrophic disaster. The December 2005 draft National Preparedness Goal defines both the 37 major capabilities that first responders should possess to prevent, protect from, respond to, and recover from a wide range of incidents and the most critical tasks associated with these capabilities. An inability to effectively perform these critical tasks would, by definition, have a detrimental impact on effective protection, prevention, response, and recovery capabilities. Since September 11, 2001, the federal government has awarded almost $14 billion in grants and assistance to state and local governments to assist in building emergency management capabilities.

Federal Government is a Major Partner in Long-Term Recovery

Because of the widespread physical and economic damage to the Gulf Coast, there are numerous examples of the federal government playing a major role in support of state and local recovery efforts. Debris removal and repairs to the region's public infrastructure and residential and commercial properties are proceeding, to a significant extent, from federal funding designated for the area and much remains to be done. The U.S. Army Corps of Engineers is reconstructing the New Orleans levee system to fix those parts of the system that failed during Hurricane Katrina. The federal role in rebuilding will be particularly important for transportation and health infrastructures and federal facilities.

Our prior work has identified state efforts underway to develop long-term rebuilding strategies. In Louisiana, the governor and the mayor of New Orleans have charged different groups with guiding various aspects of the rebuilding efforts. In Mississippi, the Governor's Commission on Recovery, Rebuilding and Renewal was formed to develop a strategy for rebuilding the affected areas of Mississippi. On November 1, 2005, the President issued Executive Order 13390, which directed the creation of a central figure in the administration's efforts to support the Gulf Coast recovery and rebuilding phases.[21] Specifically, the President directed the Secretary of Homeland Security to establish within the department the position of Coordinator of Federal Support for the Recovery and Rebuilding of the Gulf Coast region. The federal coordinator is responsible for developing principles and goals, leading the development of federal recovery activities, and monitoring the implementation of designated federal support. The coordinator also serves as the focal point for managing information flow, requests for actions, and discussions with Congress, state and local governments, the private sector, and community leaders.

Congress and the Administration Have Engaged in Numerous Efforts to Identify and Recommend Improvements to Catastrophic Disaster Preparedness, Response, and Recovery

Beginning in February 2006, several reports from Congress and the administration have explored the events surrounding Hurricane Katrina and chronicled many weaknesses and some strengths of the preparation and response efforts, providing observations and recommendations to improve national preparedness for and response to catastrophic disasters. Table 1 contains the resulting reports and a brief description of their findings.

Table 1. Findings in Reports by Congress and the Administration

Title and author	Major findings
DHS/FEMA Initial Response Hotwash: Hurricane Katrina in Louisiana (February 13, 2006) Source: FEMA	Found improvements needed in areas of communications and interoperability; FEMA staffing; unified command; logistics and staging; and operating procedures. Recommendations for FEMA included: work to strengthen emergency management capability at state and local levels; review emergency management architecture for response and recovery operations; train, equip, and staff response teams; improve the financial management of disasters; improve leadership and management; establish command authority in the Joint Field Office (JFO); and continue catastrophic planning with federal, state, and local governments.
A Failure of Initiative: Final Report of the House Select Bipartisan Committee to Investigate the Preparation for and Response to Hurricane Katrina (February 15, 2006) Source: House of Representatives House Select Bipartisan Committee	Identified 14 major findings including: critical elements of the NRP were executed late, ineffectively or not at all; there was massive communications damage; command and control was impaired at all levels, delaying relief; and the military played an invaluable role but coordination was lacking, among others. No recommendations provided.

Table 1. (Continued).

Title and author	Major findings
The Federal Response to Hurricane Katrina: Lessons Learned (February 23, 2006) Source: White House Homeland Security Council	Identified 17 critical challenges including National Preparedness, Integrated Use of Military Capabilities, Communications, Logistics and Evacuation, Search and Rescue, Public Safety and Security, Public Health and Medical Support, Human Services, Mass Care and Housing, Public Communications, Critical Infrastructure and Impact Assessment, Environmental Hazards and Debris Removal, Foreign Assistance, Non-Governmental Aid, Training, Exercises, and Lessons Learned, Professional Development and Education, Citizen and Community Preparedness. Identified 125 recommendations for DHS, DOD, and other federal departments and agencies, including 44 that were to be implemented by June 1, 2006.
A Performance Review of FEMA's Disaster Management Activities in Response to Hurricane Katrina (March 31, 2006) Source: Department of Homeland Security's Office of Inspector General	Found FEMA adapted to new response plans with difficulty; FEMA provided record levels of support but needs to improve delivery structure; and FEMA needs to improve readiness. Identified 38 recommendations.
Hurricane Katrina: A Nation Still Unprepared: Report of the Committee on Homeland Security and Governmental Affairs (May 2006) Source: U.S. Senate Committee on Homeland Security and Governmental Affairs	Found differing roles at different levels of government affected the response; long-term and short-term warnings went unheeded; preparation proved insufficient; response at all levels of government was unacceptable; long-term factors contributed to poor response; and waste, fraud and abuses were identified. Report identified seven foundational recommendations based on identified systemic weaknesses and challenges.

Source: GAO analysis of executive branch and congressional reports.

Chapter 3

Leadership is Critical for Preparing for, Responding to, and Recovering from Catastrophic Disasters

Our March 2006 testimony, along with the House, Senate, White House Homeland Security Council, and DHS reports on the preparation for and response to Hurricane Katrina discussed the vital importance of federal, state, and local leadership, including clearly defined and well understood roles and responsibilities. All the various critiques of Hurricane Katrina concluded that leadership at all levels of government should be improved. In preparing for, responding to, and recovering from any catastrophic disaster, the legal authorities, roles and responsibilities, and lines of authority for the preparation and response at all levels of government must be clearly defined, effectively communicated, and well understood in order to facilitate rapid and effective decision making. For example, during Hurricane Katrina, separate roles specified in the NRP and in the Stafford Act resulted in leadership uncertainty and may have slowed the response, and some of the NRP's provisions were unclear and may have hindered the federal government's ability to guide a more proactive response. DHS has reported taking a number of actions to address leadership issues raised in the various Hurricane Katrina reports, particularly making revisions to the NRP to clarify roles, responsibilities, and lines of authority. Another leadership issue raised in the aftermath of the disaster was whether the organizational placement of FEMA is appropriate for its mission.

Communicating and Clearly Defining Leadership Roles, Responsibilities, and Lines of Authority in Advance of Catastrophic Disasters Is Essential to Effective Response

Our March 2006 testimony, and other reports issued on the preparation for and response to Hurricanes Katrina and Rita, have all identified the importance of improved leadership and clearly defined and well understood roles, responsibilities, and lines of authority at all levels of government, especially in times of a catastrophe. In the response to Hurricane Katrina there was confusion regarding roles and responsibilities under the NRP. This included uncertainty concerning the roles of the Secretary of Homeland Security, the Principal Federal Officer (PFO), and the Federal Coordinating Officer (FCO), under the NRP.

At the time of Hurricane Katrina, there were three key roles in the management of a catastrophic disaster. First, with the passage and subsequent implementation of the Homeland Security Act of 2002, the Secretary of Homeland Security became the cabinet-level focal point for response to natural and man-made crises (such as major disasters and terrorist incidents) requiring a coordinated response and developing a single, coordinated national response plan for such crises and incidents. We view this as a strategic role to coordinate federal activities and policy from a national standpoint and be directly responsible and accountable to the President, which is consistent with recommendations we made in 1993. The revised NRP clearly states that the Secretary of Homeland Security, who reports directly to the President, is responsible for declaring and managing incidents of national significance, including catastrophic incidents.

The response to Hurricane Katrina identified unresolved issues regarding lines of authority with respect to the Secretary of Homeland Security and the FEMA Director, as well as the key officials reporting to them, the PFO to the secretary and the FCO to the FEMA Director. For example, in response to Katrina, the Secretary of Homeland Security initially designated the head of FEMA as the PFO, who appointed separate FCOs for Alabama, Louisiana, and Mississippi. It was not, however, clear who was responsible for coordinating the overall federal effort at a strategic level. Our fieldwork indicated that the lack of clarity in leadership roles and responsibilities resulted in disjointed efforts of many federal agencies involved in the response, a myriad of approaches and processes for requesting and providing assistance, and confusion about who should be advised of requests and what resources would be provided within specific time frames. The House Select Committee also found difficulties with roles and responsibilities, including federal officials' unfamiliarity with their roles and responsibilities under the NRP and NIMS. Likewise, the White House Homeland Security Council made numerous recommendations, including giving the PFO the same authority as an FCO to execute responsibilities and coordinate federal response assets. According to the White House Homeland Security Council report, giving the PFO this authority could be accomplished without a change to the Stafford Act by simply designating the PFO as an FCO.

In our March 8, 2006, testimony, we recommended that DHS clarify and communicate the roles of the secretary, the PFO, and the FCO, consistent with the provisions of the Homeland Security Act and the Stafford Act. In May 2006, DHS revised the NRP to state that the Secretary of Homeland Security may choose to combine the roles of the PFO and FCO in a single individual to help ensure synchronized federal coordination for incidents other than terrorist incidents.[22] In instances where a single PFO/FCO has been appointed, the revised NRP states that deputy FCOs will be designated for the affected states to support the PFO/FCO. In addition, DHS has pre-designated regional officials who will be PFOs and FCOs in the event of an incident of national significance.

The NRP revisions may not fully resolve the leadership issues with respect to the PFO and FCO roles. While the secretary may avoid conflicts by appointing a single individual to serve as PFO and FCO in non-terrorist incidents, confusion may persist if the secretary does not exercise this discretion. Furthermore, this discretion does not exist for terrorist incidents that may result in a Stafford Act declaration. In a terrorist-caused incident, the PFO and FCO roles cannot be combined in a single individual, as might be the case in a natural disaster. The revised NRP does not specifically provide a rationale for this limitation, although a terrorist incident would involve law enforcement and criminal investigation activities that the PFO must coordinate with the Attorney General's representative, the FBI Special Agent-in-Charge. Nevertheless, it is not clear whether, and to what extent, the PFO's roles and responsibilities might conflict with those of the FCO in a terrorist incident resulting in a Stafford Act declaration. Given the persistent confusion about the NRP regarding key federal leadership roles and responsibilities in a catastrophic disaster, we are recommending that the Secretary of Homeland Security rigorously re-test, train, and exercise its recent clarification of the roles, responsibilities, and the lines of authority for all levels of leadership, implementing changes needed to remedy identified coordination problems.

New GAO recommendation to DHS to re-test, train and exercise its recent clarification of leadership roles
DHS should rigorously re-test, train, and exercise its recent clarification of the roles, responsibilities, and lines of authority for all levels of leadership, implementing changes needed to remedy identified coordination problems.
For more detailed information about our recommendations, see appendix I.

PROVISIONS OF THE NRP REGARDING INCIDENTS OF NATIONAL SIGNIFICANCE WERE NOT CLEAR AT THE TIME OF HURRICANE KATRINA, AND REVISIONS POTENTIALLY RAISE NEW ISSUES

We identified an ambiguity in the NRP's triggering mechanisms that existed at the time Hurricane Katrina hit the Gulf Coast—i.e., what actions were necessary to activate the NRP. The NRP distinguished between incidents that required DHS coordination and those that did not. Only those requiring DHS coordination, termed "incidents of national significance," triggered activation of the NRP. However, it was not clear whether the Secretary of Homeland Security needed to make a formal declaration of such an incident to trigger the NRP or whether such an incident was automatically triggered by one of four criteria contained in Homeland Security Presidential Directive-5 (HSPD-5).[23] In our March 2006 testimony, we recommended that DHS clarify in the NRP whether the Secretary of Homeland Security must formally declare an incident of national significance to activate the NRP and, if not, whether the secretary must take any specific actions when the President, in effect, activates the NRP by declaring a Stafford Act emergency or major disaster.

The May 25, 2006, revisions to the NRP make clear that the Secretary of Homeland Security must formally declare an incident of national significance.[24] The four factors stated in HSPD-5 continue to be primary criteria for such a declaration, but the revised NRP states that the Secretary of Homeland Security is not limited to these criteria and may consider other factors in deciding whether to declare an incident of national significance. According to the revisions, the key is whether the incident is of such severity, magnitude, or complexity that it requires the Secretary of Homeland Security to manage the federal response.

While clarifying the declarations process for incidents of national significance, the revised NRP potentially raises new implementation issues. First, the May 2006 revisions of the NRP state that the NRP is "always in effect," but it is not clear what this means operationally. Previously, the NRP only applied to incidents of national significance, which were defined as those incidents requiring DHS coordination. Under the revised NRP, incidents of national significance are not simply those incidents requiring DHS coordination, but those that must be managed directly by the Secretary of Homeland Security. While the revised NRP makes clear that incidents of national significance must be declared by and managed by the Secretary of Homeland Security, the revised NRP also applies to incidents of lesser severity that may nevertheless require some federal involvement. According to the revised NRP, such incidents are to be managed by the federal department or agency with jurisdictional authority based on a scaled and flexible implementation of the NRP. There is no declarations process for incidents of lesser severity and there are no specific plans detailing how agencies are to address such incidents. In the absence of any implementation plans or a specific declaration or designation process, leadership issues may arise in responding to issues of lesser severity.

Proactive Response Provisions of the NRP's Catastrophic Incident Annex Were Unclear at the Time of Hurricane Katrina, and Implementation Issues Remain

The NRP's Catastrophic Incident Annex also was a source of considerable criticism after Hurricane Katrina. Under the NRP, a catastrophic incident is any natural or manmade incident (including terrorism) that results in extraordinary levels of mass casualties, damage, or disruption severely affecting the population, infrastructure, environment, economy, national morale, and/or government functions. A catastrophic incident could result in sustained national impacts over a prolonged period of time; almost immediately exceeds resources normally available to state, local, tribal, and private-sector authorities in the impacted area; and significantly interrupts governmental operations and emergency services to such an extent that national security could be threatened. The annex describes an accelerated, proactive national response to catastrophic incidents and establishes protocols to pre-identify and rapidly deploy essential resources expected to be urgently needed. At the time of Hurricane Katrina, a draft supplement to the annex would have limited the annex's scope to no-notice or short-notice catastrophic incidents, not incidents that may evolve or mature to catastrophic magnitude, which could be the case with strengthening hurricanes. Importantly, and consistent with a prior GAO recommendation, the revised NRP does not include this limitation. While stating that the catastrophic incident annex is primarily designed to address no-notice or short-notice catastrophic incidents that defy advance planning and resource pre-positioning, the revised NRP states that projected catastrophic incidents (e.g., hurricanes) are also covered by the annex. In the event there is time to assess the requirements and plan for a catastrophic event, the proactive federal response and pre-positioning of assets is to be tailored to address the specific situation.

A proactive approach to catastrophic disasters when there is warning is consistent with recommendations we made in 1993 following Hurricane Andrew.[25] FEMA generally agreed with our findings and recommendations and had begun taking actions to address them. We noted that some of the improvements necessary in the overall federal response were outside of FEMA's control, such as Presidential action to strengthen leadership for the federal response by designating a key White House or cabinet-level official to oversee the federal government's overall preparedness and response activities for catastrophic disasters. In 1996,

FEMA became a cabinet agency whose director reported directly to the President, consistent with our 1993 recommendation to the President.

The Homeland Security Act of 2002 made FEMA a part of DHS in March 2003, with its director reporting to the Secretary of Homeland Security. As a result of the new national response framework brought about by the Homeland Security Act, HSPD-5 and the NRP, the Secretary of Homeland Security became the cabinet-level focal point for emergency preparedness and response, and the principal Federal official for domestic incident management. The role assigned to the secretary through this framework is consistent with our 1993 recommendation. However, the experience of Katrina revealed some of the same problems that led to our 1993 recommendation, including a lack of clarity with regard to federal leadership roles and responsibilities for the response. For example, the secretary promptly designated the Director of FEMA as PFO under the NRP, whose responsibilities included serving as the secretary's personal representative on the scene, providing situational awareness, and coordinating the federal response. However, as noted in subsequent testimony and congressional reports, problems arose with regard to the FEMA Director's reporting relationship with the secretary and the White House during the response, and it was unclear who was directly accountable to the President for the overall federal strategic coordination and management of the incident. The Secretary of Homeland Security did not perform this role, and the Director of FEMA was not in a position to effectively perform this role. As noted above, the revised NRP addresses this issue with respect to incidents of national significance, by requiring the Secretary of Homeland Security to declare such incidents, including those deemed catastrophic, and to manage the federal response following such a declaration. Maintaining at the secretary level the strategic role, described earlier, of coordinating federal activities and policy from a national standpoint, and being directly responsible and accountable to the President, is critical to ensuring an efficient federal response and vital presidential leadership during a catastrophic disaster. These responsibilities should be performed by the secretary and not delegated to the PFO or FCO in such circumstances.

We continue to believe that our other 1993 recommendations still apply in terms of improving the nation's response to catastrophic disasters. For example, the NRP's catastrophic incident annex needs to be supported with more robust and detailed operational implementation plans. In addition, we are renewing our recommendation that Congress should consider giving federal agencies explicit authority to take actions to prepare for catastrophic disasters when there is warning. In 1993 we found that federal response time could be reduced by encouraging agencies to do as much advance preparation as possible prior to a disaster declaration. When there is early warning, as there is for hurricanes, DOD and other federal agencies need to mobilize resources and deploy personnel before the catastrophe strikes. However, current law—in particular, the Stafford Act—does not explicitly authorize such pre-declaration activities. Therefore, federal agencies may fail to undertake advance preparation, because of uncertainty over whether FEMA will request their assistance and whether costs incurred before a disaster declaration will ultimately be reimbursed by FEMA.

New GAO matter for congressional consideration
Congress should give federal agencies explicit authority to take actions to prepare for catastrophic disasters when there is warning.
For more detailed information about our recommendations, see appendix I.

Our review of the NRP and its catastrophic incident annex—as well as lessons from Hurricane Katrina—demonstrate the need for DHS and other federal agencies to develop robust and detailed operational plans to implement the catastrophic incident annex and its supplement in preparation for and response to future catastrophic disasters. DHS has not yet released the supplement to the catastrophic incident annex, more than a year after Katrina and its original target date. The catastrophic incident supplement is intended as the more operationally specific companion to the catastrophic incident annex. For example, the September 2005 draft supplement includes operations to be carried out by local, state, and federal responders; detailed execution schedules and implementation strategies; functional capability overviews (such as coverage for transportation support); and key responsibilities of federal departments and agencies. Until DHS finalizes and distributes the supplement, neither DHS nor other federal agencies that have responsibilities under the supplement can finalize the detailed implementation plans required to make the NRP fully operational for catastrophic disasters. Therefore, as noted in our March testimony, we are recommending that DHS direct that the NRP base plan and its Catastrophic Incident Annex be supported by more robust and detailed operational implementation plans, particularly the Catastrophic Incident Supplement to the NRP. Such operational plans should, for example, further define and leverage those military capabilities that might be needed in a catastrophic disaster.[26]

New GAO recommendation to DHS for detailed and robust implementation plans for the NRP
DHS should direct that the NRP base plan and its Catastrophic Incident Annex be supported by more robust and detailed operational implementation plans, particularly the Catastrophic Incident Supplement to the NRP. Such operational plans should, for example, further define and leverage those military capabilities that might be needed in a catastrophic disaster.
For more detailed information about our recommendations, see appendix I.

Federal Coordination Issues with the NRP Have Not Been Fully Addressed by Recent Revisions

Our ongoing Hurricane Katrina-related work has identified other examples of possible implementation issues within the NRP related to coordination within and across federal government entities. These involve coordination issues for search and rescue, the military response, worker safety and the role and responsibilities of the Red Cross. First, the NRP did not fully address search and rescue missions. While tens of thousands of people were rescued through the efforts of the military, civilian government, and private rescuers, the lack of clarity in search and rescue plans led to operations that, according to aviation officials, were not as efficient as they should have been. Under the National Search and Rescue Plan[27] the Coast Guard ordinarily had responsibility for providing or arranging maritime search and rescue services, and the Air Force ordinarily had responsibility for providing or arranging non-maritime search and rescue. The plan also called on DOD to support civil search and rescue, and it stated that DOD and Coast Guard commands should provide their facilities for civil search and rescue to the fullest extent practicable. While the NRP acknowledged the existence of the National Search and Rescue Plan, the NRP did not specifically address how the Coast Guard and the Air Force organizational responsibilities in the National Search and Rescue Plan coincided with the NRP's urban search and rescue annex. In addition, the National Search and Rescue Plan had not been updated to reflect the NRP. As a result of the lack of clear search and rescue guidance, the aviation portion of military search and rescue operations was not fully integrated with the helicopter search and rescue operations of the Coast Guard and other rescuers. Moreover, no one had the total picture of the missions that had been resourced and the missions that still needed to be performed during the response to Hurricane Katrina.

Second, the military mounted a massive response to Hurricane Katrina that saved many lives and greatly assisted recovery efforts. However, the NRP made little distinction between the military response to smaller, regional disasters and the military response to large-scale, catastrophic natural disasters even though past disasters had shown that the military tends to play a much larger role in catastrophes. We found a lack of understanding within the military and among federal, state, and local responders as to the types of assistance and capabilities that DOD might provide in the event of a catastrophe—for example, timely damage assessments or communications capabilities—the timing of this assistance, and the respective contributions of the active-duty and National Guard forces. For example, neither the NRP, DHS, nor DOD had fully identified the military's extensive reconnaissance assets or communications capabilities that could be brought to bear in a catastrophe. In the absence of this planning, some of the military's available assets were never requested or proactively deployed.

To improve the military response to catastrophic disasters, we have recommended that the Secretary of Defense (1) provide proposed revisions of the NRP to DHS that addresses the proactive functions the military is expected to make during a catastrophic incident, and (2) establish milestones and expedite the development of detailed plans and exercises to fully account for the unique capabilities and support that the military is likely to provide to civil authorities in response to the full range of domestic disasters, including catastrophic disasters. Furthermore, the plans and exercises should specifically address the use of reconnaissance, communications, and search and rescue capabilities; integration of active component and National Guard and Reserve forces; and the role the military might be expected to play in logistics.[28] DOD should direct the National Guard Bureau to work with state governors to develop and maintain a list of the types of capabilities the National Guard will likely provide in response to domestic natural disasters. DOD commented on our recommendations, partially concurring with each of them: DOD said that proactive military functions can be identified in all 15 major disaster scenarios and said it is working with the Department of Homeland Security to revise the NRP. While DOD stated that the long-term focus of the U.S. government should be to develop more robust domestic disaster capabilities within the Department of Homeland Security, it acknowledged that DOD will need to assume a more robust response role in the interim period, when other responders lack the resources and expertise to handle a particular disaster. DOD also listed a number of steps it is taking to improve its disaster response planning and exercises. It also said that consistent with its Strategy for Homeland Defense and Civil Support the active component should complement, but not duplicate, the National Guard's likely role as an early responder. DOD said that logistics planning and execution are the clear responsibility of FEMA and individual states, and DOD would remain ready in a supporting role. However, we continue to believe that DOD should plan and prepare to assume additional emergency support function responsibilities during catastrophic disasters when other responders may be overwhelmed.

Recent GAO recommendations to DOD improve the military's response to catastrophic natural disasters	DOD response
DOD should provide proposed revisions of the NRP to DHS that address the proactive functions the military is expected to perform during a catastrophic incident.	DOD is working with DHS to revise the NRP and acknowledged that DOD will need to assume a more robust response role when other responders lack the resources and expertise to handle a particular disaster.
DOD should establish milestones and expedite the development of detailed plans and exercises that address specifically the use of various military resources and role(s) the military might be expected to play in order to fully account for the unique capabilities and support that the military is likely to provide to civil authorities in response to the full range of domestic disasters, including catastrophic disasters. Plans and exercise should address the use of (1) reconnaissance, communication and search and rescue capabilities; (2) active and reserve forces; and (3) possible logistics role for the military.	DOD detailed a number of actions it is taking to enhance its disaster planning and exercises.
DOD should direct the National Guard Bureau to work with state governors to develop and maintain a list of types of capabilities the National Guard will likely provide in response to domestic natural disasters.	DOD partially concurred with this recommendation. DOD was developing scalable capability packages.
For more detailed information about our recommendations, see appendix I.	

Third, our ongoing work examining worker safety issues has found that the Department of Labor's Occupational Safety and Health Administration (OSHA) and FEMA officials disagreed about which agency was responsible for ensuring the safety and health of response and recovery workers. OSHA and FEMA disagreed over who had the leadership role as the Safety and Health Coordinator at the Joint Field Office in each state and in the field. This lack of clarity about each other's roles and procedures resulted in delayed implementation of the NRP's Worker Safety and Health Support Annex. Following a review of the draft of this report, OSHA reported that their agency and FEMA have worked together to develop procedures for role of the Safety and Health Coordinator in the Joint Field Office and for the NRP's Worker Safety and Health Support Annex.

Lastly, the Red Cross and FEMA also had differing views about their roles and responsibilities under Emergency Support Function-6 (ESF-6) of the NRP, which hampered efforts to coordinate federal mass care assistance. The two organizations differed in their understanding of the role of the ESF-6 coordinator, according to a key FEMA official tasked with providing strategic vision and leading efforts to coordinate federal mass care, housing, and human services assistance. This difference in expectations about the role of the ESF-6 coordinator created tension between FEMA and the Red Cross and affected the organizations' working relationship. Additionally, Red Cross staffing policies, which directed staff and volunteers to rotate every 2 to 3 weeks and the absence of a comprehensive FEMA system to track Red Cross requests for assistance also hindered the Red Cross and FEMA's working relationship during hurricane relief efforts. Red Cross and FEMA officials have stated they are working to clarify future roles and responsibilities.

To clarify roles and responsibilities within ESF-6 for the 2006 hurricane season, and to help ensure that FEMA's resource tracking system will meet the needs of those requesting FEMA assistance, we have recommended that the Secretary of DHS should direct FEMA to work with the Red Cross as soon as possible to reach agreement on the operating procedures that they will both use in the event of an incident of national significance and ensure that FEMA obtains input from the Red Cross as it develops a resource tracking system. DHS did not provide a response to our recommendations, noting that FEMA was actively preparing for the hurricane season. We have also recommended that the Chief Executive Officer of the Red Cross implement ESF-6 staffing strategies that better facilitate the development of working relationships and retain institutional knowledge.[29]

The Red Cross endorsed and is taking actions, as applicable, to address, two of our recommendations. For example, the Red Cross said it has been working closely with FEMA in recent months to develop and finalize a memorandum of understanding that outlines areas of mutual support and cooperation with respect to response and recovery in presidentially-declared disasters and emergencies. Overall, the Red Cross agreed with our conclusion that coordination between FEMA and the Red Cross could be improved for the 2006 hurricane season. With respect to our recommendation about staffing strategies, the Red Cross said that it is in the process of hiring ESF-6 reservists who will be deployed for extended periods of time to perform Red Cross ESF-6 mass care functions at the federal level.

Recent GAO recommendations to improve FEMA and Red Cross coordination	DHS and Red Cross response
DHS should direct FEMA to work with the Red Cross as soon as possible to reach agreement on the operating procedures that they will both use in the event of an incident of national significance.	The Red Cross commented that it has been working closely with FEMA in recent months to develop and finalize a memorandum of understanding that outlines areas of mutual support and cooperation with respect to response and recovery in presidentially-declared disasters and emergencies.
The Chief Executive Officer of the Red Cross should implement ESF-6 staffing strategies that better facilitate the development of working relationships and retain institutional knowledge.	The Red Cross said that it is in the process of hiring ESF-6 reservists who will be deployed for extended periods of time to perform Red Cross ESF-6 mass care functions at the federal level.
For more detailed information about our recommendations, see appendix I.	

DHS Reported Taking Some Actions to Improve Leadership in Response to Findings in Congress' and the Administration's Reviews

In addition to changes to the NRP, DHS has reported taking other actions to improve DHS leadership during major disasters. In July 2006, Secretary of Homeland Security Chertoff sent letters to the governors of Alabama, Louisiana, and Mississippi identifying the PFO and FCO for their respective states with the purpose of permitting coordinated, collaborative federal, state, pre-disaster planning and communications. Each of these letters differed somewhat in content, but generally summarized the agreements on DHS's assistance for each state, including pre-positioned supplies and equipment and identified key federal and state roles and responsibilities in the event of a catastrophic event. Each letter also emphasized that evacuation "is fundamentally a state and local responsibility." Similar letters were also sent to the chief executive of all the states and territories, according to DHS.

Table 2. Implementation of White House Homeland Security Council Recommendations for the 2006 Hurricane Season: Recommendations Related to Leadership Issues

Recommendation related to leadership	Primary agency(ies) responsible for implementation	Supporting agency(ies) responsible for implementation
Critical Action Recommendation: Ensure co-location of relevant federal, state, and local decision-makers, including leaders of state National Guards, to enhance unity of effort.	Department of Homeland Security	All federal agencies
Critical Action Recommendation: For events preceded by warning, ensure preparations to pre-position a fully resourced and integrated interagency Federal Joint Field Office (JFO) to coordinate and, if necessary, direct federal support to the disaster.	Department of Homeland Security	Office of Management and Budget, Department of Defense
Critical Action Recommendation: Co-locate a single Department of Defense point of contact at the JFO and current FEMA regional offices to enhance coordination of military resources supporting the response.	Department of Defense, Department of Homeland Security	N/A
Critical Action Recommendation: Identify and develop rosters of federal, state, and local government personnel who are prepared to assist in disaster relief.	Department of Homeland Security	All federal agencies
Critical Action Recommendation: Enhance ongoing review of state evacuation plans and incorporate planning for continuity-of-government to ensure continuation of essential and emergency services.	Department of Homeland Security	Homeland Security Council
Establish an interagency team of senior planners with appropriate emergency management experience to conduct a comprehensive, 90-day review of the NRP and the NIMS.	Department of Homeland Security	All federal agencies
Revise the NRP to address situations that render state and local governments incapable of an effective response.	Department of Homeland Security	All federal agencies
PFO should have the authority to execute responsibilities and coordinate federal response assets.	Department of Homeland Security	N/A
Establish National Information Requirements and a National Information Reporting Chain.	Department of Homeland Security	All federal agencies
Establish the Disaster Response Group.	Homeland Security Council	Members of the Homeland Security Council
Develop recommendations for revision of the NRP to delineate the circumstances, objectives, and limitations of when DOD might temporarily assume the lead for the federal response to a catastrophic incident.	Department of Homeland Security, Department of Defense	N/A
Revise the Immediate Response Authority policy to allow commanders, in appropriate circumstances, to exercise this authority even without a request from local officials.	Department of Defense	N/A
Establish standards for "pushing" the pre-positioning of federal assets to states and locals, in the case of an imminent catastrophe.	Department of Homeland Security	All federal agencies

Source: GAO analysis of DHS and Homeland Security Council data .

Note: N/A = not applicable.

DHS has also reported taking other actions to improve DHS leadership during major disasters through communications such as press releases, speeches by top officials, and congressional testimonies by the Secretary of Homeland Security, the Undersecretary for Preparedness, the recently confirmed Undersecretary for Federal Emergency Management (who is also the FEMA Director), and FEMA's Director of Operations. These reported actions include improvements to national preparedness and integrating the use of the military's capabilities. Table 2 lists the White House Homeland Security Council recommendations regarding leadership, with the White House Homeland Security Council report's "critical action" recommendations. According to DHS officials, they were to have implemented 11 recommendations that the White House Homeland Security Council said were essential for the 2006 hurricane season. However, DHS has not provided us with the documentation needed to verify these actions. DHS officials have stated that they have put a more experienced senior leadership team in place at FEMA and created a National Operations Center to improve coordination and situational awareness.

As a Result of FEMA's Performance During Hurricane Katrina, Questions Have Been Raised about Its Organizational Placement

FEMA's performance problems in responding to Hurricane Katrina have been well documented in the several reports. In addition to the problems related to roles and responsibilities and lines of authority previously discussed, the House Select Committee noted that (1) senior officials were ill prepared due to their lack of experience and knowledge of the required roles and responsibilities prescribed by the NRP; (2) having lost a number of top disaster specialists, senior leaders, and other experienced personnel since 2002, DHS and FEMA lacked adequately trained and experienced staff for the Katrina response; and (3) FEMA's logistics systems were unable to support large-scale logistical challenges. There is concern among members of Congress that the causes of FEMA's response and recovery problems may stem from FEMA's organizational placement and its budgetary relationship within DHS.

Different approaches have been suggested for addressing FEMA's performance problems in responding to Hurricane Katrina. The White House Homeland Security Council report's recommendations covered areas such as improving DHS expertise and experience, development of a national crisis communications system, and development of DHS regions that are fully staffed, trained, and equipped to manage and coordinate all preparedness activities and any emergency that may require a substantial federal response. There are some proposals in Congress that would keep FEMA within DHS, but statutorily reorganize the agency's responsibilities and lines of authority. Other proposals would remove FEMA from DHS and reconstitute it as an independent agency, much as it was prior to the creation of DHS. As we stated in our March 8 and May 9, 2006, testimonies, organizational changes alone, while potentially important, are not enough to adequately address the underlying systemic conditions that resulted in FEMA's performance problems. In our view, a number of factors may be ultimately more important to FEMA's success in responding to and recovering from future disasters than its organizational placement. Conditions underlying FEMA's performance during Hurricane Katrina involved the experience and training of DHS or FEMA leadership; the clarity of FEMA's mission and related responsibilities and authorities to achieve mission performance expectations; the adequacy of its human, financial, and technological resources; and the effectiveness of planning, exercises, and related partnerships. These issues must be addressed whether or not FEMA remains in DHS. For example, we believe that the position of FEMA Director and other key FEMA leadership and managerial positions could benefit from having statutory professional qualifications requirements. In addition, Congress should consider a term appointment for the FEMA Director and other selected positions.

If an organizational change remains under consideration, our past work could be helpful. Before the formation of DHS in 2003, we testified before the House Select Committee on Homeland Security that reorganizations of government agencies frequently encounter start-up problems and unanticipated consequences, are unlikely to fully overcome obstacles and challenges, and may require additional modifications in the future.[30] Some considerations from our prior work that are relevant to FEMA's organizational placement include.

- *Mission Relevancy:* Is homeland security a major part of the agency or program mission? Is it the primary mission of the agency or program?
- *Similar Goal and Objectives:* Does the agency or program being considered share primary goals and objectives with the other components and programs in the department as a whole?
- *Leverage Effectiveness:* Does the agency or program under consideration create synergy and help to leverage the effectiveness of other agencies and programs within the department as a whole?
- *Gains through Consolidation:* Does the agency or program being considered improve the efficiency and effectiveness of homeland security missions through eliminating duplications and overlaps, closing gaps, and aligning or merging common roles and responsibilities?

In addition, the dispersion of responsibility for preparedness and response across more than one federal agency was a problem we identified during the formation of DHS. When Congress created DHS, it separated FEMA's responsibilities for preparedness and response activities into two DHS directorates. Responsibility for preparedness for terrorism disasters was placed in the department's Border and Transportation Security Directorate, which included FEMA's Office of National Preparedness. Other types of FEMA disaster preparedness and response efforts were transferred to the department's Emergency Preparedness and Response Directorate, which included FEMA. In January 2003, we observed that this organizational arrangement would challenge FEMA in ensuring the effective coordination of preparedness and response efforts and enhancing the provision and management of disaster assistance for efficient and effective response. A division of responsibility remains under the recent DHS reorganization resulting from Secretary Chertoff's Second Stage Review. Preparedness efforts—including planning, training, exercising, and funding—are consolidated into a Preparedness Directorate, while FEMA maintains responsibility for response and recovery missions and reports directly to the Secretary of Homeland Security.[31] Secretary Chertoff has stated that the reorganization would refocus FEMA on its historic mission of response and recovery. Legislation has been introduced in both the House and Senate that would reorganize the emergency management structure of the federal government. Two of the bills, for example, would amend the Homeland Security Act to create a new emergency management organization within DHS that would combine FEMA and the Preparedness Directorate.[32]

Chapter 4

Enhanced Capabilities for Catastrophic Response and Recovery Are Needed

Developing the capabilities needed for large-scale disasters is part of an overall national preparedness effort that is designed to integrate and define what needs to be done, where, based on what standards, how it should be done, and how well it should be done. The nation's experience with Hurricane Katrina reinforces some of the concerns surrounding the adequacy of the nation's capabilities in the context of a catastrophic disaster—particularly for capabilities such as the assessment of the disaster's effects, communications, logistics of supplies and services, and mass care and sheltering of victims. Ensuring that needed capabilities are available requires effective planning and coordination and training and exercises whereby capabilities are realistically tested, and problems are identified and subsequently addressed across all stakeholders. DHS has reported taking some actions to improve capabilities in response to findings in Congress' and the administration's reviews. However, ongoing work is still needed by DHS to address FEMA's significant human resource challenges (e.g., a large number of open positions, skill gaps, and succession planning challenges). Finally, risk management principles can be applied as part of the development of capabilities to guide federal, state, and local decision making by thinking about risk systematically within the normal cycle of development and implementation at all levels.

Capabilities Were Not Adequate for a Catastrophic Disaster

The various reports and our own work on FEMA's performance before, during, and after Hurricane Katrina suggest that FEMA's human, financial, and technological resources and capabilities were insufficient to meet the challenges posed by the unprecedented degree of damage and the resulting number of hurricane victims. For example, the Senate's report concluded that FEMA did not have the resources to fulfill the mission and respond effectively in a catastrophic event and recommended that DHS develop the national capabilities—especially adequate surge capacity—it needs to respond to catastrophic disasters, ensuring it has sufficient full-time staff, response teams, contracting personnel, and adequately trained and sufficiently staffed reserve corps to ramp up capabilities, when needed. The Senate report also identified the need for DHS to complete and/or adopt technology and information management systems to effectively manage catastrophic disaster-related activities. The report concluded that resources are needed for staffing and preparation of regional strike teams, better development of a trained cadre of reservists, and the development of new logistics capabilities. FEMA's assessment of its initial response concluded that the agency needed to lead an audit of current staffing capability and workforce demands for staff in a severe or catastrophic event and determine the number of personnel available to serve in each position or unit for such an event. Their assessment also concluded that FEMA needs to develop a communications suite that operates independently of normal communications infrastructure and is able to be moved into disaster locations. Similarly, the White House Homeland Security Council report identified the need for each homeland security region to have access to the resources, equipment, and personnel needed to establish a self-sufficient, temporary Joint Field Office to direct response and recovery efforts anywhere within the region.

In developing its lessons learned from Hurricane Katrina, the White House Homeland Security Council has recommended that future preparedness of the federal, state, and local authorities should be based on the risk, capabilities, and needs structure of the National Preparedness Goal. More specifically, the White House Homeland Security Council recommended that the National Preparedness Goal and its target capabilities list should be used, for example, in (1) defining required capabilities and what levels of those capabilities are needed, including those within the purview of the federal government; (2) developing mutual aid agreements and compacts informed by the National Preparedness Goal's requirements; and (3) developing strategies to meet required levels of capabilities that prioritize investments on the basis of risk.

Developing the capabilities needed for large-scale disasters is part of an overall national preparedness effort that is designed to integrate and define what needs to be done, where, based on what standards, how it should be done, and how well it should be done. The NRP defines "preparedness" as the "range of deliberate, critical tasks and activities necessary to build, sustain, and improve the operational capability to prevent, protect against, respond to, and recover from domestic incidents. Preparedness is a continuous process involving efforts at all levels of government and between government and private-sector and nongovernmental organizations to identify threats, determine vulnerabilities, and identify required resources." In our earlier work on the National Preparedness Goal, we observed that if properly planned and executed, the goal and its related products, such as program implementation plans and requirements, may help guide the development of realistic budget and resource plans for an all-hazards national preparedness program.[33] However, questions remain regarding what should be expected in terms of the expanded capabilities and mutual aid needed from other jurisdictions to meet the demands of a catastrophic disaster. The nation's experience with Hurricane Katrina reinforces some of the questions surrounding the adequacy of capabilities in the context of a catastrophic disaster—particularly in the areas of (1) situational assessment and awareness, (2) emergency communications, (3) evacuations, (4) search and rescue, (5) logistics, and (6) mass care and sheltering.

Situational Assessment and Awareness

Situational assessment and awareness activities are a critical capability common across all phases of an incident (i.e., preparedness, response, and recovery), to immediately estimate services needed by communities and disaster victims. The NRP notes that following a disaster, state and local governments are responsible for conducting initial damage assessments, but it also notes that state and local officials could be overwhelmed in a catastrophe. Our prior work has pointed out that the NRP does not specify the proactive means or capabilities the federal government should use to conduct damage assessments and gain situational awareness when the responsible state and local officials are overwhelmed. For example, the military has significant assets to provide situational assessment and awareness, and although some of its capabilities were employed during Hurricane Katrina, there had been no advance planning among federal, state, and local responders as to how DOD might support this capability in the event of a catastrophic disaster. As a result, response efforts were hampered by the federal government's failure to fully use its available assets to conduct timely, comprehensive damage assessments in Louisiana and Mississippi. In 1993, we also identified the failure to quickly assess damage and gain situational awareness as a problem during Hurricane Andrew in 1992.[34] We recommended then that FEMA improve its catastrophic disaster response capability by using existing authority to aggressively respond to catastrophic disasters, assessing the extent of damage, and then advising state and local officials of identified needs and the federal resources available to address them.

Our analysis shows that neither the NRP nor DOD's functional plan specifically called for the proactive use of the military's extensive reconnaissance assets as part of a proactive response to catastrophic incidents. About 4 days after Katrina's landfall the military began providing imagery data from some of its reconnaissance assets to its forces and civilian agencies, although some information was classified due to its source and could not be shared directly with civilian agencies. Also, some agencies were not able to access some of the available information because the data files were too large to download. A National Guard Hurricane Katrina after-action review reported that the adjutants general in Mississippi and Louisiana required real-time imagery that the military community should have been able to provide, but did not. Because state and local officials were overwhelmed and the military's extensive reconnaissance capabilities were not effectively leveraged, responders began organizing and deploying without fully understanding the extent of the damage or the required assistance. In contrast, DOD officials told us that almost immediately after Hurricane Rita struck Louisiana and Texas in September 2005, considerable assets were made available to assess damage, primarily because of lessons learned from Hurricane Katrina.

To improve the military response to catastrophic disasters, we have recommended that the Secretary of Defense establish milestones and expedite the development of detailed plans and exercises to fully account for the unique capabilities and support that the military is likely to provide to civil authorities in response to the full range of domestic disasters, including catastrophes; and that the plans and exercises should specifically address the use of reconnaissance capabilities.[35] In May 2006, the Assistant Secretary of Defense for Homeland Defense testified that U.S. Northern Command was developing a reconnaissance annex to its Defense Support to Civil Authorities Contingency Plan and had tested the annex's concepts during an 11-day exercise in May 2006.

Recent GAO recommendations to DOD to improve the military response to catastrophic disasters	DOD response
DOD establish milestones and expedite the development of detailed plans and exercises capabilities and support to civil authorities in response to the full range of domestic disasters, including catastrophes; the plans and exercises should specifically address the use of reconnaissance capabilities.	U.S. Northern Command was developing a reconnaissance annex to its Defense Support to Civil Authorities Contingency Plan and had tested the annex's concepts during an 11-day exercise in May 2006.
For more detailed information about our recommendations, see appendix I.	

Emergency Communications

Emergency communications is a critical capability common across all phases of an incident. Agencies communications systems during a catastrophic disaster must first be operable, with sufficient communications to meet everyday internal and emergency communication requirements. Once operable, they then should have communications interoperability whereby public safety agencies (e.g., police, fire, emergency medical services, etc.) and service agencies (e.g., public works, transportation, and hospitals) can communicate within and across agencies and jurisdictions in real time as needed. Hurricane Katrina caused significant damage to the communication infrastructure in Louisiana and Mississippi, which further contributed to a lack of situational awareness for military and civilian officials. Hurricane Katrina destroyed or severely degraded many commercial landline and cellular telephone systems. In addition, emergency radio systems usage exceeded their capacity, making it difficult to establish necessary connections between officials and responders at the local, state, and federal levels, and to implement other capabilities. For example, even when local officials were able to conduct damage assessments, the lack of communications assets caused delays in transmitting the results of the assessments. As a result, it was difficult for officials to share situational awareness.

The military, other government agencies, and public companies all have extensive communications assets and capabilities, but DHS has responsibility for coordinating communications for disaster response operations under the NRP. Our work has found, however, that neither DHS nor DOD fully identified the extensive military communication capabilities that could be leveraged as part of a proactive federal response to catastrophic disasters.[36] For example, although DOD's emergency response plan addressed internal military communications requirements, DOD did not address the communication requirements of communities affected by a catastrophic natural disaster, and it did not address coordination with civilian responders. As previously noted, we have made recommendations to the Secretary of Defense to develop detailed plans and exercises regarding the unique capabilities and support, such as communications capabilities, that the military is likely to provide to civil authorities.[37] In May 2006, the Assistant Secretary of Defense for Homeland Defense testified that DOD had taken a number of steps to increase communications capabilities in regions affected by disasters. These include authorizing the Command of U.S. Northern Command to deploy communications capabilities and develop pre-scripted requests for assistance for deployable communications options.

Recent GAO recommendations to DOD improve federal leadership and cooperation in communications	DOD response
DOD should develop detailed plans and exercises regarding the unique capabilities and support, such as communications capabilities, that the military is likely to provide to civil authorities.	DOD has taken a number of steps to increase communications capabilities in regions affected by disasters. These include authorizing the Command of U.S. Northern Command to deploy communications capabilities and develop pre-scripted requests for assistance for deployable communications options.
For more detailed information about our recommendations, see appendix I.	

Our prior work on interoperable communications identified problem definition, performance goals and standards, and defining the roles of federal, state, local government and other entities as the three principal challenges to achieving effective interoperable communications for first responders. The single greatest barrier to addressing the decades-old problems of interoperable communications has been the lack of effective, collaborative, interdisciplinary, and intergovernmental planning. Our 2004 recommendations called for DHS to: (1) work with the Federal Communications Commission (FCC) to develop a nationwide database of interoperable communications frequencies and a common nomenclature so that first responders from different disciplines and jurisdictions can quickly identify shared frequencies when arriving at the scene of an incident; (2) establish interoperability requirements whose achievement can be measured; and (3) encourage states through DHS grants to establish statewide bodies that will develop a comprehensive statewide interoperable communications plan and condition the purchase of new equipment on the adoption of such a plan.[38] DHS responded that it was developing a nationwide database of interoperable public safety communications frequencies and also said it planned to work on a common nomenclature across public safety disciplines and jurisdictions. DHS also said it is developing a methodology to establish a national baseline of public safety communication and interoperability capabilities with input from the public safety community. On June 16, 2006, the FCC published a Notice of Proposed Rulemaking to address and implement the recommendations of an independent panel that reviewed the impact of Hurricane Katrina on communications networks. The FCC identified 18 actions that the Commission can take in conjunction with the private sector and state and local governments, and other federal departments, to promote heightened readiness and preparedness, and effective and efficient response and recovery efforts.[39]

Evacuations

By definition, a catastrophic disaster like Hurricane Katrina will impact a large geographic area, necessitating the evacuation of people—including vulnerable populations, such as hospital patients, nursing home residents, and transportation-disadvantaged populations who were not in such facilities. Although state and local governments can order evacuations, health care facilities can be exempt from these orders. Hospital and nursing home administrators often face challenges related to evacuations caused by hurricanes, including deciding whether to evacuate and obtaining transportation. Nursing home administrators must locate receiving facilities that can accommodate residents who may need a place to live for a long period of time. If a hospital or nursing home does decide to evacuate, contractors providing transportation for hospitals and nursing homes could be unable to provide enough vehicles during a major disaster because local demand for transportation could exceed supply.

We have conducted work related to the evacuation of special needs populations in the aftermath of Hurricane Katrina. The National Disaster Medical System (NDMS) is the primary federal program that supports the evacuation of patients in need of hospital care during disasters such as hurricanes. We found that the program has two limitations in its design that constrain its assistance to state and local governments with patient evacuation. The first limitation is that NDMS evacuation efforts begin at a mobilization center, such as an airport, and do not include short-distance transportation assets, such as ambulances or helicopters, to move patients out of health care facilities to mobilization centers. Even during a catastrophe when state and local government capabilities are almost immediately overwhelmed, short-distance transportation remains a state and local responsibility.[40] The second limitation is that NDMS supports the evacuation of patients needing hospital care; the program was not designed nor is it currently configured to move people who do not require hospitalization, such as nursing home residents, which remains a state and local responsibility.

NDMS supplemented state and local emergency response capabilities with federal resources and services and helped evacuate about 2,900 people during recent hurricanes, including Hurricane Katrina. According to program officials the program was actually designed to evacuate and transport hospital patients starting from a federally-designated mobilization center, although NDMS supported evacuation efforts during Hurricane Katrina that included nursing home residents. NDMS officials explained that the program does not have agreements with nursing homes that could receive evacuated nursing home residents or appropriate transportation such as special buses that could transport them.

To address limitations in how the federal government provides assistance with the evacuation of health care facilities, we have recommended that the Secretary of Homeland Security clearly delineate how the federal government will assist state and local governments with the movement of patients and residents out of hospitals and nursing homes to a mobilization center where NDMS transportation begins; and in consultation with the other NDMS federal partners—the Secretaries of Defense, Health and Human Services, and Veterans Affairs—clearly delineate how to address the needs of nursing home residents during evacuations, including the arrangements necessary to relocate these residents.[41] DHS said it would take the recommendation under advisement as it reviews the National Response Plan, confirming that the federal government becomes involved in evacuations only when the capabilities of state and local governments are overwhelmed.

Recent GAO recommendations to DHS to improve federal assistance for health facility and nursing home disaster evacuations	DHS response
DHS should clearly delineate how the federal government will assist state and local governments with the movement of patients and residents out of hospitals and nursing homes to a mobilization center where NDMS transportation begins.	DHS said it would take the recommendation under advisement as it reviews the NRP and stated that all NDMS federal partners are currently reviewing a memorandum of agreement with a view towards working with state and local partners to delineate and clarify roles and responsibilities.

DHS should in consultation with the other NDMS federal partners—the Secretaries of Defense, Health and Human Services, and Veterans Affairs—clearly delineate how to address the needs of nursing home residents during evacuations, including the arrangements necessary to relocate these residents.	
For more detailed information about our recommendations, see appendix I.	

We conducted work examining the nation's efforts to protect children after the Gulf Coast hurricanes and identified evacuation challenges for this population.[42] Thousands of children were reported missing to the National Center for Missing and Exploited Children, which utilized its trained investigators to help locate missing children after the evacuation.[43] Officials from this center stated that both the American Red Cross and FEMA had some information on the location of children in their databases; however, they said it was difficult to obtain this information because of privacy concerns. These officials told us that standing agreements for data sharing among organizations tracking missing children, the Red Cross, and FEMA could help locate missing persons more quickly. Regarding foster children, Louisiana child welfare officials told us it was difficult to contact foster parents who had evacuated because the emergency contact information for them was limited and located in case records that was inaccessible for weeks following the storm. They said updated emergency contact information for foster parents and automated case file systems could help locate and serve foster children more quickly. Also, child welfare officials told us that having an adequate number of trained staff present during large-scale evacuations can help ensure that families stay together.

Search and Rescue

Search and rescue is the capability to coordinate and conduct urban search and rescue response efforts for all hazards. Urban search and rescue response efforts require a seamless transition from finding stranded people, rescuing them, and transporting them to safe shelter, which may call on the additional skills and resources of responders, along with support from other emergency support functions such as public safety. Capabilities also can include accessing, medically stabilizing, and extricating victims trapped in damaged or collapsed structures. The U. S. Coast Guard, state and local agencies, and military assets rescued thousands in the aftermath of Hurricane Katrina. Almost 6,000 Coast Guard personnel from throughout the country conducted one of the largest search and rescue missions in its history as part of an even larger multi-agency, multi-level search and rescue effort, according to the White House Homeland Security Council report. The Coast Guard retrieved more than 33,000 people along the Gulf Coast, including more than 12,000 by air, and 11,000 by surface, plus 9,403 evacuated from hospitals. Almost one-third of the Coast Guard's entire fleet was dedicated to rescue efforts, according to the White House Homeland Security Council report.

Search and rescue capabilities must also include ensuring the safety of first responders. For example, following Hurricane Katrina's landfall, Coast Guard rescue swimmers involved in the helicopter-based rescues told us that safety and security became a concern as time passed and individuals became frustrated because they had no food or water. In some instances, tensions among survivors became heated when rescue swimmers prioritized the rescues of children, women, and the elderly over able-bodied men. For boat operations, officials explained that they dealt with security challenges by deploying Coast Guard security teams with the rescue personnel to provide armed security coverage for both Coast Guard personnel and FEMA urban search and rescue teams.

In addition, although tens of thousands of people were rescued after Hurricane Katrina through the efforts of military, civil government, and private rescuers, the lack of clarity in search and rescue plans led to operations that were not as efficient as they should have been. As we noted in our discussion of leadership, the NRP at that time addressed only part of the search and rescue mission, and the National Search and Rescue Plan had not been updated to reflect the NRP. According to Louisiana National Guard officials, they worked with the Coast Guard to coordinate aviation operations, but the aviation search and rescue efforts that were being directed from two different command sites were not integrated. For example, some military aircraft received their direction from a military task force at the Superdome, while others received their direction from the Coast Guard or the Joint Task Force that was in command of the active troops on the ground. According to military officials, better integration of search and rescue efforts could reduce duplications of effort for search and rescue aircraft. As previously noted, GAO has made recommendations to the Secretary of Defense to improve planning for the search and rescue support that the military is likely to provide to civil authorities.[44]

Logistics

Our work and that of others indicates that logistics systems—the capability to identify, dispatch, mobilize, and demobilize and to accurately track and record available critical resources throughout all incident management phases—-were often totally overwhelmed by Hurricane Katrina. Critical resources apparently were not available, properly distributed, or provided in a timely manner. The result was duplication of deliveries, lost supplies, or supplies never being ordered. Reviews of acquisition efforts indicated that while these efforts were noteworthy given the scope of Hurricane Katrina, agencies needed additional capabilities to (1) adequately anticipate requirements for needed goods and services; (2) clearly communicate responsibilities across agencies and jurisdictions; and (3) deploy sufficient numbers of personnel to provide contractor oversight. For example, a factor that affected the military response was the large and unanticipated logistics role it was asked to assume. Under the NRP, FEMA is responsible for coordinating logistics during disaster response efforts, but during Hurricane Katrina, FEMA quickly became overwhelmed, in part because it lacked the people, processes, and technology to maintain visibility—from order through final delivery—of the supplies and commodities it had ordered. As a result of FEMA's lack of visibility over the meals that were in transit, DOD had to airlift 1.7 million meals to Mississippi to respond to a request from the Adjutant General of Mississippi, who was concerned that food supplies were nearly exhausted.

Similarly, our work examining the coordination between FEMA and the Red Cross to provide relief to disaster victims found that FEMA did not have a comprehensive system to track requests for assistance it received from the Red Cross on behalf of voluntary organizations and state and local governments for items such as water, food, and cots. The absence of such a system created more work for the Red Cross and slowed the delivery of relief services. The Red Cross was only able to follow up on these requests informally—a process that took time and was often ineffective. FEMA officials were often unable to provide the Red Cross with accurate information regarding FEMA's ability to fulfill a request or when expected items would be delivered, causing many requests to go unfilled or be filled too late to be of use. The unreliability of FEMA's supply systems required the Red Cross to try to follow up on requests through other informal channels—a process the Red Cross reported as being inefficient and only marginally effective. Other voluntary organizations also told us that in many cases the unreliability of FEMA's supply systems hindered their attempts to provide mass care services, and, as a result, they had to scale back on their service provision.

In contrast, response efforts benefited from the logistical support of the U.S. Forest Service. As the White House Homeland Security Council reported in February 2006, about 3,000 members of the Forest Service also deployed to the region to support response efforts. Forest Service personnel in Mississippi, Louisiana, Florida, and Alabama established support camps, provided aviation assistance, and transported needed supplies to relief workers. The base camps they established were capable of supporting 1,000 emergency responders at each site. According to the Forest Service, nearly 2,760 Forest Service employees nationwide were called upon to provide assistance by using their incident management abilities to help manage evacuation centers and base camps, provide logistical support, clear roadways, and operate mobilization centers and trailer staging areas. They also helped navigate the federal procurement system and successfully obtained needed emergency response supplies, the White House Homeland Security Council noted. Forest Services' incident management teams helped to supply more than 600,000 people with 2.7 million meals, 4 million gallons of water, and 40 million pounds of ice. One Forest Service region sent nearly 470 people, including various types of interagency responder teams, buying teams, and other services in the relief effort, according to the Forest Service.

Mass Care, Housing, and Human Services

This emergency support function includes the non-medical mass care, housing, and human services needs of individuals and families. Mass care is the capability to provide immediate shelter, feeding centers, basic first aid, and bulk distribution of needed items and related services to affected persons. Charities and government agencies that provide human services, supported by federal resources, helped meet the mass care needs of the hundreds of thousands of evacuees.

Recognizing the historically large role of charities in responding to disasters, DHS collaborated with the American Red Cross in the development of the NRP and gave it considerable responsibilities. In addition, the NRP establishes a separate support annex that defines the capability to effectively manage and deploy volunteers and unsolicited donations. Federal and charitable organization officials we spoke to indicated that because of the catastrophic nature of the storms, in some cases volunteers and donations were not well integrated into response and recovery activities.

Charities have taken steps to improve coordination of relief efforts since September 11, 2001. These efforts have included sharing information about relief services through daily conference calls and the use of electronic databases. Additionally, charities coordinated service delivery to meet the needs of evacuees. For example, the Red Cross did not work in areas that might become flooded or in structures that could be compromised by strong winds. During our visits to the Gulf Coast region in October 2005, we observed that in areas where the Red Cross did not provide services, the Salvation Army and smaller organizations—often local churches—were able to meet many of the charitable needs of hard-to-reach communities. Despite these efforts, charities faced some challenges in coordinating service delivery. For example, some charities reported that the electronic database created to share information about services provided to evacuees was still in its developmental stage and, therefore, not ready to be activated on such a large scale. Some concerns were also raised about smaller charities' abilities to provide adequate disaster relief services. Additionally, some Gulf Coast areas did not have sufficient shelter facilities, in part because the Red Cross had in effect "shelter protection policies" to promote evacuation of low-lying areas, which prohibited the placement of shelters in areas prone to high winds, storm surge, and flooding—causing some of the most severely affected areas to be without sheltering facilities. In addition, some shelters were accessible to the general population only during limited daylight hours.

In addition to immediate food and shelter, individuals may need help replacing Social Security checks, obtaining food stamps or other basic federal support benefits and services. Our preliminary work indicated that a number of federal and state programs provided assistance and services to eligible individuals and families before the catastrophic disaster. Significant ongoing assistance after the catastrophic disaster has also been required. Such basic federal support programs as food stamps, social security, unemployment insurance, and income assistance played an important role in the immediate response to those displaced or otherwise affected by Hurricane Katrina. To ensure delivery of benefits, and in some cases, to address the sharp rise in new applications, these programs set up operations in shelters, disaster recovery centers, other sites, and some programs served an unprecedented number of people in their offices that serve the public. For example, Hurricane Katrina disrupted mail delivery of monthly social security checks, leaving many social security recipients without key income support. However, the Social Security Administration had procedures in place to provide emergency payments and, with about 1,300 offices nationwide, had the capability to quickly deploy staff and equipment from other offices to address the increased workload. Individuals affected by Hurricane Katrina also overwhelmed the state of Louisiana's food stamp, welfare, and unemployment offices. State and local officials we visited said they struggled to find equipment and supplies to handle the increase and to obtain other help such as security for crowd control. Louisiana—overwhelmed in its efforts to issue unemployment checks to evacuees without stable addresses—rushed to implement a debit card payment system and encountered some challenges doing so. The debit card payment systems the states of Alabama, Louisiana, and Mississippi had in place for income assistance and food stamps allowed them to continue benefit payments that were uninterrupted when beneficiaries relocated. In addition, over 2 million families received food assistance through the U.S. Department of Agriculture's Disaster Food Stamp Program.

Effective Capabilities Development Requires Adequate Planning, Coordination, Training, and Exercises

Capabilities are built upon the appropriate combination of people, skills, processes, and assets. Ensuring that needed capabilities are available requires effective planning and coordination and training and exercises in which the capabilities are realistically tested, problems identified and lessons learned, and subsequently addressed in partnership with other federal, state, and local stakeholders.

Planning and Coordination

As we identified in our March 2006 testimony, the ability of the nation to prepare for, respond to, and recover from catastrophic disasters can be enhanced through strong advance planning, both within and among responder organizations. By their very nature, catastrophic disasters involve extraordinary levels of mass casualties, damage, or disruption that likely will immediately overwhelm state and local responders, circumstances that make sound planning for catastrophic events all the more crucial. Identifying, obtaining, and pre-positioning adequate quantities of key supplies, such as food, ice, water, and blankets, and delivering those supplies quickly when and where needed is an important component of planning. Another example is the previously discussed planning challenges states and localities face in preparing for and carrying out the evacuation of transportation-disadvantaged populations, such as the elderly or persons with disabilities, during a disaster. Our work in this area has noted challenges in identifying these populations, determining their needs, and providing for and coordinating their transportation. Some emergency management officials told us they did not yet have a good understanding of the size, location, and composition of the transportation-disadvantaged in their communities. However, we have also observed efforts in some locations to address the evacuation needs of the transportation-disadvantaged by encouraging citizens to voluntarily register with their local emergency management agency, integrating social service providers into emergency planning, and other measures.

Examples of Good Planning

In some cases, sound advance planning contributed to a more effective response during Hurricane Katrina. For example:

- U. S. Coast Guard–The Coast Guard was able to mitigate some of the communication shortfalls it experienced, in part because of its planning assumption that "communications systems could be heavily damaged or destroyed" during a natural disaster. This assumption prompted Coast Guard officials to build into their contingency plans approaches that were not reliant on communication systems and that allowed personnel to act independently or with limited guidance from commanding officers. As a result of the contingency plan, personnel in charge of the assets knew their mission prior to the storm and did not need to communicate any further with district command to fulfill their operation. In addition, as planned, the Coast Guard pre-placed communication equipment before the storm. Members of the Coast Guard Auxiliary were notified prior to the hurricane's landfall and provided communication capabilities after the storm passed, which according to Coast Guard officials, was critical to conducting search and rescue operations.

- National Finance Center—Planning for continuity of operations enabled the National Finance Center (NFC) in New Orleans paid more than half a million federal workers on time while Hurricane Katrina was bearing down, then made subsequent payrolls—NFC's largest ever—without any delays. According to officials, NFC had gone through various diverse scenarios in disaster recovery planning and exercises as part of its preparedness for just such an event. There was a structured timeline for reviewing the continuity of operations plan every year, beginning with continuity of operations plan requirements and business impact analysis at the unit level. A NFC official said that key NFC staff members worked the weekend before landfall at the New Orleans facility to complete payroll processing for federal employees, then shut down operations and deployed to backup locations before the hurricane hit New Orleans. NFC backup data was trucked out of the New Orleans facility. When Hurricane Katrina made landfall in Louisiana shortly after 6 a.m. on Monday, August 29, 2005 NFC had already sent an advance deployment team to its backup sites in Philadelphia, Pennsylvania and Grand Prairie, Texas. That same night, the trucks arrived at an alternate processing facility outside Philadelphia. The backup tapes also enabled the New Orleans NFC staff restore many operations within 2 days of the devastation. Within a 50-hour period, NFC had the data center recovered and began catch-up processing to bring applications up to current state.
- Social Security Administration---The Social Security Administration had enhanced planning and pre-established procedures in place to provide immediate emergency payments to the significant number of beneficiaries who evacuated and did not receive their monthly checks. With these procedures in place, the Social Security Administration had the capability to deploy staff and equipment from its 1,300 offices across the nation to address the increased workload.
- Department of Defense—The White House Homeland Security Council's report also identified DOD's preparedness efforts for the 2005 hurricane season. DOD approved a standing order to prepare and organize for severe weather disaster operations, which was based on prior assistance for hurricane recovery operations. This order expedited the pre-positioning of senior military representatives known as Defense Coordinating Officers, to act as liaisons with other governmental organizations in the projected disaster area prior to an event. The order also authorized the use of DOD installations as logistical staging areas for FEMA. The military also took steps to proactively respond as Katrina strengthened in the Gulf of Mexico, and it published warning and planning orders and was positioned to respond with both National Guard and federal forces by the time Katrina made landfall on August 29, 2005.
- U.S. Postal Service—The Postal Service was relatively well-prepared for the aftermath of Hurricane Katrina. While the hurricane damaged some 500 post offices, the Postal Service was able to establish limited services just days after the disaster. As early as August 26, 2005, USPS was anticipating a landfall in the Gulf Coast and was diverting some mail from processing plants in that region to minimize the effects

of potential disruption to its processing capabilities. By the time the storm passed on August 30, some 500 postal facilities in Alabama, Mississippi, and Louisiana had been damaged and were incapable of providing delivery service. On September 1, 2005, the Postal Service set up temporary centers for the pickup of checks in Mississippi, Louisiana, and Alabama. In New Orleans proper, two post offices were designated as Social Security check pickup points for a short period of time. Reconstituting regular mail delivery in the areas affected by Hurricane Katrina was a priority. For those now displaced from their homes, the Postal Service says that it can provide relatively efficient delivery service if customers submit a change of address form in person, by toll-free telephone, or on the Internet. The Postal Service now has the capability, through innovations in sorting technology, to intercept mail at its original sorting point that is addressed to an address that has changed, and re-route the mail to a new address. Thus, it can avoid the cost and delay of physically transporting mail to the former delivery post office for flagging and forwarding under a local change of address request.

- Internal Revenue Service—The Internal Revenue Service (IRS) worked in coordination with FEMA to address the human services needs of individuals and families. The IRS began staffing Disaster Assistance Centers in more than a dozen states. The IRS assigned nearly 5,000 employees to augment the FEMA hurricane victim registration effort and established its own dedicated toll-free disaster number and a special section of their internet web site, according to the Treasury Department. IRS employees were taking calls seven days a week over two shifts to help people with the process of registering with FEMA to obtain benefits, answering approximately 950,000 registration calls for FEMA and filling orders for over 291,000 Disaster Relief Kits. Through February 2, 2006, they answered more than 100,000 calls on the special IRS toll-free line for affected taxpayers. In addition, the IRS issued filing and payment relief guidelines, launched a special toll-free disaster hotline, and created a Hurricane Katrina disaster link at *www.irs.gov* web site to expedite the processing of transcript and tax return requests.

Examples Where Planning and Coordination Should Be Improved

- DHS Nationwide Plan Review: State and Local Catastrophic Disaster Planning—In the aftermath of Hurricane Katrina, DHS has responded to the need to ascertain the status of the nation's emergency preparedness planning, as identified both by Congress and the President. On September 15, 2005, the President ordered DHS to undertake an immediate review, in cooperation with local counterparts, of emergency plans in every major city in America. In addition, the Safe, Accountable, Flexible, Efficient Transportation Equity Act: A Legacy for Users required the Secretary of Transportation and the Secretary of Homeland Security to jointly review and assess federal and state evacuation plans for catastrophic hurricanes impacting the Gulf Coast Region.[45] Finally, in the conference report to the Department of Homeland Security Fiscal Year 2006 Appropriations Act, the conferees directed DHS to report on the status of catastrophic planning, including mass evacuation planning, in all 50 states and the 75 largest urban areas.[46] In response, DHS developed the

Nationwide Plan Review in coordination with the Department of Transportation and support from the Department of Defense.

The Nationwide Plan Review included two phases. In Phase 1, released in February 2006, all states and urban areas submitted self-assessments of their emergency operations plans, focusing on their adequacy and feasibility to manage the consequences of a catastrophic event.[47] In Phase 2, released in June 2006, DHS employed peer review teams to visit states and urban areas, where these teams reviewed and validated the self-assessments, and helped determine requirements for federal planning assistance.

At the conclusion of each visit, the peer review team completed a comprehensive report and submitted it to DHS.

The June 2006, DHS Phase 2 report found that current catastrophic planning is unsystematic, not linked within a national planning system, and the status of plans and planning gives grounds for significant national concern. The report found a systemic problem of outmoded planning processes, products, and tools, which all contribute to inadequate catastrophic planning. DHS found that emergency operations rely on plans that are created in isolation, insufficiently detailed, and not subject to adequate review. The report identified 15 findings specific to state and urban area planning, and 24 findings specific to the federal government's role and efforts to support catastrophic disaster planning. DHS concluded that the result of these systemic planning problems translates to uneven performance and repeated and costly operational miscues, and critical response time lost to correct the misperceptions of federal, state, and local responders about their roles, responsibilities, and actions.

DHS's findings call for a fundamental modernization of the nation's planning processes. According to its report, planning modernization must be managed as a single program with established funding. Further, according to DHS, the goal of the modernization program must be to establish a networked, collaborative national planning system that satisfies planners' information needs; provides procedures and tools to accomplish pre-incident plan synchronization; allows faster development or revision of existing plans; and provides flexible options that accommodate the diverse hazards and threats. However, DHS states that the next phase that needs to be addressed is the development of an implementation strategy to begin to address the 40 findings in the White House Homeland Security Council report in a rational way. Such an implementation strategy can prioritize remedial actions based on a risk management model that considers threats, vulnerabilities, and consequences.

- Department of Defense—Our recent report on DOD and the National Guard's planning for and response to Hurricane Katrina illustrates an example of the need for coordinated planning in advance of a disaster. While multiple agencies support each "Emergency Support Function" in the NRP, DOD does not have the primary responsibility for any emergency support function, as DOD's role is primarily that of a support agency. We found that pre-Katrina plans involving the military were inadequate in several ways. Neither the NRP nor DOD's disaster plans incorporate lessons learned from past catastrophes to fully delineate the military capabilities needed to respond to a catastrophic disaster.[48] DOD is aware of disaster response problems described in this report and is beginning to take actions to address the lessons learned from Hurricane Katrina and to prepare for the next catastrophic event. DOD has been conducting its own reviews and is also examining the lessons and recommendations in reports from a White House review panel, congressional oversight committees, and other sources. We noted that DOD is taking some actions to address catastrophic disaster response problems. For example, DOD officials stated that it is currently updating its emergency response plan and intends to use a contingency plan rather than a less detailed functional plan to guide its military support to civil authority missions. DOD also has an organizational realignment underway that gives a single Army organization that is responsible for domestic disaster response and will be capable of deploying within 18 hours as joint task forces for catastrophes anywhere in the United States.

- FEMA—Enhancing planning and coordination efforts may also benefit other federal agencies' efforts to provide response and recovery services. For example, our observations of the Individuals and Households Program (IHP), which provides housing and other financial assistance to disaster victims, suggest that FEMA's program planning was reactive and ad hoc, with IHP procedures differing from disaster area to disaster area. Our work examining the IHP program indicated that FEMA lacked final plans, policies, and procedures that specifically addressed the types of unique challenges the agency could expect to face in catastrophic circumstances. In commenting on a draft of our report, FEMA officials said they had previously identified the need to develop plans to expand its disaster registration intake and applicant assistance process and to temporarily relocate victims to outside the area after a catastrophic disaster. However, they said the $20 million to address these requirements and others that were requested and provided in the FY 2005 Disaster Supplemental appropriation were not available for commitment until days before Katrina made landfall. They also said that their planning efforts were significantly slowed by staff commitments to the 2004 and 2005 hurricane seasons. One of the consequences of this lack of planning for catastrophic disasters was that FEMA's systems and processes for verifying applicant eligibility for IHP assistance were overwhelmed, resulting in payments made to thousands of ineligible applicants.[49] FEMA officials said that these problems more directly reflected a failure of system capacity that could not be overcome by any amount of planning. Nonetheless, if FEMA had developed and implemented a plan for an expandable disaster registration intake and applicant assistance process after identifying the need for such a plan, systems and processes for verifying applicant eligibility for IHP assistance would have been better prepared to manage the unprecedented volume of requests and reduce the number of payments made to ineligible applicants.

In addition, FEMA did not have sufficient pre-positioned supplies, equipment, and services (e.g., debris removal) contracts in place prior to Katrina to meet the demand during the response, particularly for needs such as temporary housing and public buildings. The experience of Katrina highlighted the need for better logistics planning and the need for contracts to be in place prior to the disaster that could be activated to lean forward and provide surge capacity for critical supplies and services. For example, because FEMA had not planned for a catastrophic disaster such as Katrina, FEMA spent funds for temporary housing that were hardly used. In addition, because FEMA did not inform the Corps of Engineers prior to Katrina that the Corps would be responsible for acquiring portable classrooms, the Corps lacked sufficient knowledge of the market for this commodity to obtain the best deal for the government. In these and other instances, better planning could have avoided some unnecessary costs.

- Small Business Administration—In another example, SBA needs enhanced strategic planning to ensure its ability to provide timely loans, as our ongoing work in this area indicates. As the primary federal lender to disaster victims SBA's disaster loan program provides loans to businesses, homeowners, and renters to rebuild and replace uninsured or underinsured property damaged by a disaster. However, as more fully described in our recently issued report, several factors affected SBA's ability to provide timely disaster assistance to victims of the Gulf Coast hurricanes, including the large volume of applications that SBA processed.[50] For example, SBA planned the maximum user capacity for its new Disaster Credit Management System (DCMS) based solely on its experience during the 1994 Northridge earthquake—the single largest disaster SBA had previously faced—and other historical data. SBA did not consider information available from catastrophe risk models and disaster simulations, such as the likelihood and severity of damages from potential natural disasters, to help predict the volume of applications that might be expected from such events. SBA's limited planning contributed to insufficient DCMS user capacity, thus restricting the number of staff that could access DCMS and process applications in a timely manner. SBA also did not completely stress test DCMS before implementation and received the incorrect computer hardware from its contractor, which reduced user capacity and contributed to the system instability, outages, and slow response times initially experienced by SBA staff. As a result of these and other factors, SBA faced significant delays and backlogs in processing loan applications.

Going forward, SBA may be able to process disaster loans more efficiently by implementing an Internet-based application feature. In order to provide more timely disaster assistance in the future, we recommended that the Administrator of SBA direct the Office of Disaster Assistance to take the following four actions: (1) reassess DCMS's maximum user capacity and related loan processing resource needs based on such things as lessons learned from the Gulf Coast hurricanes, a review of information available from catastrophe risk modeling firms and disaster simulations, and related cost considerations; (2) conduct complete stress testing to ensure that DCMS can function as planned for maximum user capacity levels; (3) improve management controls over assessing contractor performance through inspections of all equipment purchased or leased to support DCMS; and (4) expedite plans to resume business processing reengineering efforts to analyze the disaster loan process and identify ways to more efficiently process loan applications, including an evaluation of the feasibility of implementing a secure Internet-based application feature for home loan applicants. SBA disagreed with some of our findings but generally agreed with these recommendations.

In addition our preliminary analysis from an ongoing review indicates that SBA's overall planning efforts for providing timely recovery assistance in response to large scaled disasters appears to be insufficient and may have contributed to the delays in providing affordable disaster assistance to victims in the Gulf Coast Region. For example, at the time of the Gulf Coast hurricanes, SBA lacked a comprehensive, documented plan for providing timely disaster recovery assistance. Disaster recovery experts have told us that such a plan would address various aspects of SBA's response including staffing capacity, telecommunication needs and other logistical support, as well as coordination with federal, state, and local entities in providing appropriate disaster recovery assistance. In addition, SBA did not have any full-time staff responsible for planning its disaster recovery activities. Rather, SBA relied extensively on the experience and knowledge of its field office staff to determine its response needs and carry out SBA's role. SBA officials stated that it recently started developing a response plan; however, it was unclear when this plan would be completed and implemented. We plan to complete our review of these issues and issue a report later this year.

Recent GAO recommendations to SBA to provide more timely disaster assistance in the future	SBA response
SBA should reassess the Disaster Credit Management System's (DCMS) maximum user capacity and related loan processing resource needs based on such things as lessons learned from the Gulf Coast hurricanes, a review of information available from catastrophe risk modeling firms and disaster simulations, and related cost considerations.	SBA disagreed with some of our findings but generally agreed with these recommendations.
Conduct complete stress testing to ensure that DCMS can function at planned for maximum user capacity levels.	
Improve management controls over assessing contractor performance through inspections of all equipment purchased or leased to support DCMS.	
Expedite plans to resume business processing reengineering efforts to analyze the disaster loan process and identify ways to more efficiently process loan applications including an evaluation of the feasibility of implementing a secure Internet-based application feature for home loan applicants.	
For more detailed information about our recommendations, see appendix I.	

- Child Welfare—Findings from our recent work and from other agency "after action" reports point to the need to improve current plans and strengthen related agreements and understandings of expectations between governments and other organizations that will be responders in the event of a future catastrophe. Our work has identified areas where prior to an incident, obstacles can be identified and agreements to address them reached between agencies, levels of governments, and private and nongovernmental organizations. Our recent work examining Louisiana's child welfare system's response to the disaster and education system identified lessons learned by the state's child welfare officials, including the need for state disaster plans to include evacuation information and instructions for social workers and their foster parents. In a recent national survey, child welfare officials in 20 states and the District of Columbia reported that they had a written child welfare disaster plan.[51] Of these, 13 states reported that their plan addressed identifying children under state care who may be dispersed after a disaster, and 14 reported that their plan addressed continuing services for children under state care who may be dispersed. Child welfare agencies could also benefit from standing data-sharing agreements that would speed efforts to locate displaced children in the event of the type of evacuation witnessed prior to and after Katrina's landfall.[52] Louisiana child welfare officials told us that, during the response to Hurricane Katrina, they had to sign a memorandum of understanding for sharing information with the American Red Cross, but by the time the memorandum was finalized and approved the demand had abated, and the Red Cross had closed its shelters.

To better assist states in developing child welfare disaster plans, we recommended to the Secretary of Health and Human Services that child welfare disaster planning guidance should address the dispersion of children and families within and across state lines. This guidance should include information on (1) preserving child welfare records, (2) identifying children who may be dispersed, (3) identifying new child welfare cases and providing services, (4) coordinating services and sharing information with other states, and (5) placing children from other states. We also recommended that the secretary develop and provide training on child welfare disaster planning to all states. HHS responded by stating that it has taken action to update the guidance and provide training to states and will encourage them to develop and submit disaster plans for review. Finally, to ensure continuity of services within or across state lines for the children under state care, we have recommended that Congress should consider requiring that states develop and submit child welfare disaster plans for HHS review.

Recent GAO recommendations to DHHS and the Congress to safeguard children in states' child welfare systems	HHS responses
DHHS should ensure that the department's child welfare disaster planning guidance address the dispersion of children and families within and across state lines, and develop and provide training on child welfare disaster planning to all states.	HHS has taken action to update the guidance and provide training to states and will encourage them to develop and submit disaster plans for review.
To ensure continuity of services within or across state lines for the children under state care, Congress should consider requiring that states develop and submit child welfare disaster plans for HHS review.	
For more detailed information about our recommendations, see appendix I.	

Training and Exercises

Clear roles and coordinated planning are necessary, but not sufficient by themselves to ensure effective disaster management. It is important to test the plans and participants' operational understanding of their roles and responsibilities through robust training and exercise programs. Training and exercising are designed to test emergency management plans and increase the level of understanding of those roles and responsibilities on the part of officials, contrasting catastrophic versus non-catastrophic disasters. Involving key federal, state, and local leaders—including elected leaders—in robust training and exercise programs can better familiarize and prepare leaders with their roles in a catastrophic disaster. DHS has taken the lead to establish and maintain a comprehensive training and exercise program and standards to meet the national preparedness goal, as required by Homeland Security Presidential Directive 8 (HSPD-8). For example, the 15 national planning scenarios developed by DHS and the Homeland Security Council provide the basis for disaster exercises throughout the nation.

In our previous work on Hurricanes Andrew and Hugo,[53] we identified the need for the federal government to upgrade training and exercises for state and local governments specifically geared towards catastrophic disaster response. Hurricane Katrina demonstrated the potential benefits of applying lessons learned from training exercises and experiences with actual hurricanes, as well as the dangers of ignoring them. During our fieldwork, we found examples of how an incomplete understanding of NRP and NIMS roles and responsibilities could lead to misunderstandings, problems, and delays. In Louisiana, for example, some city officials were unclear about federal roles. In Mississippi, we were told that county and city officials were not implementing NIMS because they did not understand its provisions.

Our recent review of DOD's preparedness and response to Hurricane Katrina reported that inadequate exercises prior to Hurricane Katrina created a lack of understanding within the military and among federal, state, and local responders as to the types of assistance and capabilities, the timing of assistance, and the contributions that the military might provide. We noted that DOD is likely to contribute substantial support to state and local authorities, including search and rescue assets, evacuation assistance, provision of supplies, damage assessment assets, and possibly helping to ensure public safety. However, we found that few exercises led by DHS or DOD focused on catastrophic natural disasters and none called for a major deployment of DOD capabilities.

In the aftermath of Hurricane Katrina, we have made several recommendations designed to build the capabilities to respond to and recover from catastrophic disasters. To ensure that agencies are adequately prepared to continue performing essential functions following an emergency, we recommended to DHS that it improve the assessment and oversight of agency continuity planning and develop guidance on telework in such planning.[54] DHS partially agreed and stated that FEMA will be conducting assessments in conjunction with its upcoming interagency exercise. To improve the military response to catastrophic disasters, as previously noted, we recently called for improving military plans and exercises and resolving response problems associated with damage assessment, communication, search and rescue, and logistics issues.[55] We noted that DOD is taking steps to improve its future response, including expanding its training programs to accommodate planners from other agencies and to improve its disaster response planning and exercises. Also as first noted in our March 8, 2006, testimony, we are recommending that documents such as the NRP and the catastrophic incident annex be supported and supplemented by more detailed and robust operational implementation plans. Such operational plans should, for example, further define and leverage any military capabilities as might be needed in a catastrophic disaster.

Recent GAO recommendations to DHS to improve federal continuity of operations	DHS responses
DHS should conduct an assessment of agency continuity of operations plans.	FEMA will be conducting assessments in conjunction with its upcoming interagency exercise.
DHS should develop a methodology for assessing agency continuity of operations plans.	DHS stated it had a methodology, but GAO believes it will not fully ensure emergency preparedness.
DHS should establish a timeline for developing agency guidance for telework.	FEMA will coordinate with OPM to develop further telework guidance.
For more detailed information about our recommendations, see appendix I.	

Finally, as initially presented in our March 8, 2006, testimony, we are recommending that DHS should provide guidance and direction for federal, state, and local planning, training, and exercises to ensure such activities fully support preparedness, response, and recovery responsibilities at a jurisdictional and regional basis. This should also include the application of lessons learned from actual catastrophes and other disasters.

New GAO recommendations to DHS improve federal, state, and local training and exercises
DHS should provide guidance and direction for federal, state, and local planning, training, and exercises to ensure such activities fully support preparedness, response, and recovery responsibilities at a jurisdictional and regional basis. This should also include the application of lessons learned from actual catastrophes and other disasters.
For more detailed information about our recommendations, see appendix I.

In addition, we observed in our earlier work on the National Preparedness Goal that if properly planned and executed, the goal and its related products, such as program implementation plans and requirements, may help guide the development of realistic budget and resource plans for an all-hazards national preparedness program.[56] However, questions remain regarding what should be expected in terms of basic capabilities for most disasters compared to the expanded capabilities and mutual aid needed from other jurisdictions to meet the demands of a catastrophic disaster. Consequently, we recommend that DHS take the lead in monitoring federal agencies' efforts to meet their responsibilities under the NRP and the interim National Preparedness Goal, including the development, testing, and exercising of agency operational plans to implement their responsibilities under the NRP, NIMS, and the National Preparedness Goal.

New GAO recommendations to DHS monitor federal agencies' efforts to meet their responsibilities under the NRP
DHS should take the lead in monitoring federal agencies' efforts to meet their responsibilities under the NRP and the interim National Preparedness Goal, including the development, testing, and exercising of agency operational plans to implement their responsibilities under the NRP, NIMS, and the National Preparedness Goal.
For more detailed information about our recommendations, see appendix I.

DHS Reported Taking Some Actions to Improve Capabilities in Response to Findings in Congress' and the Administration's Reviews

DHS and the administration have engaged in an effort to respond to the White House Homeland Security Council's study of lessons learned about capabilities from Hurricane Katrina. The White House Homeland Security Council's report originally identified 125 recommendations to improve future disaster preparedness, response, and recovery capabilities. As table 3 shows, 27 recommendations that were to be implemented prior to June 1, 2006, are focused on improving capabilities. However, DHS has not provided information to support the actions it has reported taking to implement these 27 recommendations, including which actions have resulted in changes in operational readiness and capabilities.

Table 3. Implementation of White House Homeland Security Council Recommendations for the 2006 Hurricane Season: Recommendations Related to Regarding Capabilities

Recommendation related to capabilities ("Critical Action" recommendations are noted)	Primary agency(ies) responsible for implementation	Supporting agency(ies) responsible for implementation
Critical Action Recommendation: Establish rapid deployable communications as well as institute a structure for consolidated federal operational reporting to the Department of Homeland Security to ensure situational awareness	Department of Homeland Security	Department of Defense, Department of Commerce, Federal Communications Commission
Critical Action Recommendation: Encourage states to pre-contract with service providers for key disaster relief efforts, such as debris removal and the provision of critical commodities.	Department of Homeland Security	Department of Justice, Office of Management and Budget
Critical Action Recommendation: Designate locations throughout the country for receiving, staging, moving, and integrating federal disaster relief personnel and assets to ensure the most effective employment.	Department of Homeland Security	All federal agencies
Critical Action Recommendation: Update and utilize the national Emergency Alert System in order to provide the general public with advanced notification of and instruction for disasters and emergencies.	Department of Homeland Security	Department of Defense, Department of Commerce, Federal Communications Commission, Department of Health and Human Services
Critical Action Recommendation: Enhance the mechanism for providing federal funds to states for preparations upon warning of an imminent emergency.	Department of Homeland Security	Office of Management & Budget, Department of Treasury, Homeland Security Council
Establish a National Information and Knowledge Management System.	Department of Homeland Security	All federal agencies
Establish a National Reporting System.	Department of Homeland Security	All federal agencies
Promote future preparedness of the federal, state, and local authorities based on the risk, capabilities and needs structure of the National Preparedness Goal.	Department of Homeland Security	All federal agencies
Establish standard that each Homeland Security Region must be able to establish a self-sufficient, initial JFO anywhere within the region.	Department of Homeland Security	All federal agencies
Complete the review of National Security and Emergency Preparedness (NS/EP) communications policy by April 30, 2006.	Department of Homeland Security	All federal agencies
Organize an interagency group to begin the development of a national emergency communications strategy. An interim strategy to be completed May 31, 2006, should provide sufficient guidance and direction to address the deficiencies identified in the Hurricane Katrina response.	Homeland Security Council, Office of Science and Technology Policy	All federal agencies
Establish and maintain a deployable communications capability to quickly gain and retain situational awareness when responding to catastrophic incidents.	Department of Homeland Security	N/A
Establish a Chief Logistics Officer to oversee all logistics operations across multiple support functions.	Department of Homeland Security	N/A
Streamline procedures for issuing mission assignments to other departments and agencies. These mission assignments will be identified in advance of an emergency so that logisticians can operationalize assets and provide resource support rapidly. In addition, other departments and agencies should establish procedures for promptly executing mission assignments.	Department of Homeland Security	All federal agencies

Table 3. (Continued).

Recommendation related to capabilities ("Critical Action" recommendations are noted)	Primary agency(ies) responsible for implementation	Supporting agency(ies) responsible for implementation
Designate DOT as the primary federal agency responsible for developing the federal government's capability to conduct mass evacuations when disasters overwhelm state and local governments. DOT should, in coordination with HHS, DOD, Veterans Affairs, DHS, and the American Red Cross (ARC) plan, train, and conduct exercises for the timely evacuation of patients and transportation of medical supplies and personnel.	Department of Transportation	All federal agencies
Evaluate (in coordination with the Department of Transportation) all state evacuation plans as well as the evacuation plans of the 75 largest urban areas.	Department of Homeland Security	Department of Transportation
Develop the capability to surge federal law enforcement resources in the immediate aftermath of a disaster.	Department of Justice	All federal agencies
DOJ and DHS should each develop, in coordination with the other, the capability to rapidly deploy a contingent of federal law enforcement officers to prevent and respond to civil disorder.	Department of Justice, Department of Homeland Security	All federal agencies
Develop a unified and strengthened public health and medical command for federal disaster response.	Department of Health and Human Services	All Emergency Support Function (ESF)-8 Agencies
Establish a rapidly-deployable public affairs teams, able to operate self-sufficiently, in austere conditions. These teams should be established across all federal departments and agencies with key Homeland Security responsibilities.	Department of Homeland Security	All federal agencies
Establish an integrated public alert and warning system in coordination with all relevant departments and agencies.	Department of Homeland Security	All federal agencies
Review, revise, and finalize the Interim NIPP within 90 days.	Department of Homeland Security	All federal agencies
Form an Impact Assessment Working Group to provide an overall economic impact assessment of major disasters, including the Departments of Homeland Security, Treasury, Commerce, Energy (Energy Information Administration), and Labor as well as the President's Council of Economic Advisors.	National Economic Council	Department of Homeland Security, Treasury, Commerce, Energy ,Labor, President's Council of Economic Advisors
Establish an office with responsibility for integrating non-governmental and other volunteer resources into federal, state, and local emergency response plans and mutual aid agreements. Further, DHS should establish a distinct organizational element to assist faith-based organizations.	Department of Homeland Security	N/A
Conduct state and local officials training and exercises.	Department of Homeland Security	All federal agencies
Establish a National Exercise and Evaluation Program (NEEP).	Department of Homeland Security	All federal agencies
Develop the capability to rapidly activate a JTF-State for contingencies.	Department of Defense	N/A
Provide training, technical, and other assistance in support of other departments' and agencies' homeland security professional development programs.	Department of Homeland Security	All federal agencies

Source: GAO analysis of DHS and Homeland Security Council data.

Note: N/A = not applicable.

ADDITIONAL WORK IS NEEDED TO ADDRESS FEMA'S HUMAN RESOURCE CHALLENGES

The various reports and our own work on FEMA's performance before, during, and after Hurricane Katrina suggest that FEMA's human resources were insufficient to meet the challenges posed by the unprecedented degree of damage and the resulting number of hurricane victims. The Senate's report concluded that FEMA did not have the resources necessary to fulfill the mission and respond effectively in a catastrophic event and recommended that DHS develop the national capabilities—especially surge capacity—it needs to respond to catastrophic disasters, ensuring it has sufficient full-time staff and the support, such as contracting personnel and adequately trained and sufficiently staffed reserve corps, to ramp up capabilities, as needed. FEMA's initial response assessment concluded that the agency needed to lead an audit of current staffing capability and workforce demands for staff in a severe or catastrophic event and determine the number of personnel available to serve in each position or unit for such an event.

As stated during the March 8, 2006 hearing, strategic national leadership is provided by the Secretary of Homeland Security, who is to act as a focal point for natural and manmade crises and emergency planning under the provisions of the Homeland Security Act. We stated our belief that other strategic national leadership positions such as the Undersecretary for Federal Emergency Management (who is also the Director of FEMA) and other key leadership and managerial positions within FEMA and DHS could benefit from having statutory, professional qualifications requirements. In addition, Congress could consider a term appointment for the Undersecretary for Federal Emergency Management and selected other positions within DHS.

To improve the staffing level of its permanent full-time employees, on February 22, 2006, FEMA announced a hiring initiative to fill identified vacancies and attain a 95 percent staffing level in 95 days. The identified vacancies are based on the positions FEMA had already allocated but not staffed; however, the current allocation of positions has not been validated in a strategic workforce planning process. This staffing effort was to be completed before the hurricane season that began on June 1. However, the hiring initiative did not meet this target. Although FEMA was able to increase its staffing level from 76.5 percent to 81.6 percent between February 22 and May 29, the agency did not reach a staffing level of 95 percent. In June, a FEMA Human Resources official told us that while Human Resources was working toward reaching the 95 percent staffing goal in August, the agency has not established a "hard" deadline because of the priority FEMA gives to life-saving and life-sustaining operations in emergencies that could significantly slow or suspend other operations. Although not part of the 95-percent staffing effort, a related effort to fill Senior Executive Service vacancies has not made similar progress, decreasing from 74 percent to 61 percent between February 22 and May 29. According to FEMA officials, they expected that the time required to fill Senior Executive Service positions would be greater than the time required to fill other positions. In addition, the number of FEMA Senior Executive Service positions has increased since February 22, thus FEMA's level of Senior Executive Service staffing on May 29 was less than at the beginning of the period. Finally, as of May 29, FEMA had six Senior Executive Service positions that, although not vacant, had incumbents who were in an "acting" capacity.

Building Capabilities for Catastrophic Disasters Calls for a Risk Management Approach

In earlier work that included an examination of the draft National Preparedness Goal, we observed that if properly planned and executed, the goal and its related products, such as program implementation plans and requirements, may help guide the development of realistic budget and resource plans for an all-hazards national preparedness program. However, questions remain regarding what should be expected in terms of basic capabilities for most disasters compared to the expanded capabilities and mutual aid needed from other jurisdictions to meet the demands of a catastrophic disaster.[57]

Risk management can be central to assessing needs and resources in the event of future catastrophic disasters. A decision-making process using risk-management principles can guide the development of federal capabilities and the expertise that can be used to respond effectively to catastrophic disasters. The goal of risk management is to integrate systematic concern for risk into the normal cycle of agency decision making and implementation. In the aftermath of Hurricane Katrina, we identified the need for a risk management decision-making approach to develop the nation's capabilities and expertise to respond to a catastrophic disaster. Given the likely costs, Congress may wish to consider the use of a risk-management framework as it carries out its oversight and legislative responsibilities with regard to national preparedness and the recovery.

Our risk management framework calls for risk assessment based on three components of risk—threat, vulnerability, and criticality (i.e., the severity of the consequences of an incident). Using the information gathered through risk assessment, agencies can then: (1) assess the likelihood that an adverse event will occur whether caused by nature or man; (2) identify and select among alternatives to reduce the vulnerability to such an event; and (3) take actions that might reduce the consequences of one or more potential adverse events should they occur. This is a continuous process and should consider all hazards. Our risk management framework calls for links between risk-mitigation strategies, strategic goals, and plans and budgets. The framework also calls for an assessment of the value and risks of various courses of action as a tool for setting priorities and allocating resources, and the use of performance measures to assess outcomes and adjust future actions as needed.

Building and sustaining needed capabilities should be based on a risk assessment that would identify the vulnerabilities of communities or regions to potential disasters of various magnitudes and causes and how they should be addressed within available resources and with contingency planning. Periodic assessments of all hazards should determine if plans remain viable, actual capabilities match planned capabilities, and the contingency plans are appropriate. Because different states and areas face different risks, not every state or area should be expected to have the same capability to prepare for a catastrophic disaster. In our previous work examining the draft National Preparedness Goal, we observed that DHS's assessment and reporting implementation plan, intended to accurately identify the status of capabilities at the state, regional, and local levels, is vital for establishing a baseline and providing an ongoing feedback loop upon which preparedness decisions at multiple levels of government can be based. Assessment of catastrophic disaster planning and capability needs will be a critical piece. As a result, we recommend that DHS use an all-hazards, risk management approach in deciding whether and how to invest federal resources in specific capabilities for a catastrophic disaster.

New GAO recommendation to DHS to use a risk management approach
Given that resources are finite, DHS should use a risk management approach in deciding whether and how to invest in specific capabilities for a catastrophic disaster.
For more detailed information about our recommendations, see appendix I.

Chapter 5

BALANCE NEEDED BETWEEN QUICK PROVISION OF ASSISTANCE AND ENSURING ACCOUNTABILITY TO PROTECT AGAINST WASTE, FRAUD, AND ABUSE

Effective controls and accountability mechanisms for the use of resources during a catastrophic disaster are essential to ensure that resources are used appropriately, but there is a tension between normal controls and accountability procedures and the need to deliver assistance expeditiously. As we stated in February 2006, the inspectors general of the various federal departments have been conducting much of the detailed Hurricane Katrina-related work on fraud, waste, and abuse in individual federal programs.[58] In the wake of Hurricane Katrina, contracts were let quickly with little competition, funds were spent for temporary housing that was never used, and hundreds of millions of dollars might have been provided to ineligible persons for cash assistance because of an inability to verify their social security numbers or other information. Our work, and that of the DHS Inspector General and others across the audit community, have identified a number of problems that have resulted in inadequate accountability for the use of resources during and after Hurricane Katrina, including (1) contracting problems; (2) accounting for and managing international assistance that was offered; (3) tracking and managing food, water, and ice deliveries; and (4) screening fraudulent applications for assistance. These problems might have been avoided if there had been better planning for delivering assistance during catastrophic disasters.

CATASTROPHIC DISASTER CONTROLS AND ACCOUNTABILITY MECHANISMS SHOULD BE FLEXIBLE TO ADDRESS THE URGENT NEED FOR DELIVERY OF ASSISTANCE

Ensuring accountability during a catastrophic disaster requires the appropriate people, processes, and technology to, among other things, qualify benefit applicants; report and account for contracts awarded (in advance of or after the event) for such things as debris removal, temporary shelter, feeding, and medical care for disaster victims; and documenting reimbursable expenses, such as first responder overtime. Of necessity, initial response efforts focus on the immediate, urgent tasks of search and rescue and services such as medical care and food and shelter for those displaced by the disaster. In any major disaster there is the difficult task of putting in place controls and accountability mechanisms that reduce the potential for waste, fraud, and abuse but are flexible enough to provide assistance and resources quickly.

To date, Congress has appropriated approximately $88 billion of federal support through emergency supplemental appropriations to federal agencies for hurricane disaster relief and recovery efforts related to the 2005 hurricanes. [59] For Stafford Act activities, Congress makes appropriations to the Disaster Relief Fund, which FEMA administers. The Stafford Act, the principal federal disaster assistance statute, authorizes three general types of major disaster assistance: (1) public assistance grants to state and local governments and certain private nonprofit organizations, (2) hazard mitigation grant programs, and (3) individual assistance programs. For example, in the individual assistance area, FEMA may provide direct assistance (temporary housing units) and financial assistance (grant funding for temporary housing and other disaster-related needs) to disaster victims through IHP. FEMA had provided about $5.6 billon in IHP benefits as of April 2006.

As we stated in our March 8, 2006, testimony, catastrophic disasters not only require a different magnitude of capabilities and resources for effective response, but they may also require more flexible policies and operating procedures. In a catastrophe, streamlining, simplifying, and expediting decision making should quickly replace "business as usual" and the unquestioned following of long-standing policies and operating procedures used in normal situations for providing relief to disaster victims. When there is a catastrophic disaster, temporarily suspending certain rules and regulations may be necessary in order to expedite relief and recovery of the affected area, even if such a suspension requires legislation. The key is to recognize when flexibility is needed to meet response and recovery needs in a catastrophic disaster. Across our work we have discovered many examples where quick action could not occur as agencies followed procedures that required extensive, time-consuming processes, delaying the delivery of vital supplies and other assistance. In other cases, urgent need was used to bypass standard procedures without better outcomes. Processes and controls must be sufficient to provide the documentation needed for expense reimbursement and reasonable assurance that resources have been used legally and for the purposes intended.

As we saw in the aftermath of Hurricane Katrina, the lack of internal controls and other accountability mechanisms established prior to the disaster can result in delay, uncertainty, and wasted or misdirected resources. The aftermath of a catastrophic disaster is not the time to determine what rules and procedures to suspend or streamline. It is important that essential accountability mechanisms be designed and ready to implement prior to the event, just as an incident management structure should be understood and ready to implement prior to an event. Decentralization of responsibilities enhances the challenge of assuring that controls and accountability mechanisms are being followed.

Hurricane Katrina Highlighted Several Contracting Deficiencies

The government's response to Hurricanes Katrina and Rita depended heavily on contractors to deliver ice, water, and food supplies; patch rooftops; and provide housing to displaced residents and temporary facilities to local government agencies. Audits by the inspectors general at several agencies, along with the major Hurricane Katrina "after action" reports, identified deficiencies in the award and execution of many of the individual contracts. From a broader perspective, our past work has shown that, to ensure successful acquisition outcomes in any environment, certain critical success factors must be in place: sound acquisition planning, good business arrangements, and effective contract monitoring and oversight. To do so, decision makers and acquisition personnel need sufficient knowledge and clearly defined and communicated roles and responsibilities. We identified deficiencies under each of the critical success factors.

Acquisition Planning

Our work on contracting issues following Hurricane Katrina indicates that some key federal agencies involved in responding to the disaster did not have adequate acquisition plans for carrying out their assigned responsibilities. For example, while contracts for some items were in place prior to the storm, FEMA did not adequately anticipate needs for such services as providing temporary housing and public buildings. Better planning for requirements could have avoided some costs, such as the $3 million FEMA spent for 4,000 base camp beds that were never used. Another example was the government's approach to acquiring refrigerated truck services. During the 2005 hurricane season, the U. S. Army Corps of Engineers (Corps), the Department of Transportation, and FEMA had contracts with three separate contractors for refrigerated truck services.[60] We did not find a coordinated plan for obtaining these services. By not taking a coordinated approach, the agencies may have missed opportunities to make the most cost-effective use of their contracts. For example, the government obtained selected refrigerated truck services using an existing contract with Department of Transportation at an average price over 60 percent higher than the price FEMA paid for similar services.

Business Arrangements

Having good business arrangements requires that agencies have sufficient knowledge about the goods and services available in the market to satisfy their requirements, as well as the ability to translate that knowledge into sound business decisions in using their contracts. We found instances where insufficient knowledge of the market or unsound ordering practices led to excessive or wasteful expenditures.

In one case, FEMA tasked the Corps with acquiring temporary classrooms for Mississippi within a very short time frame. To meet the requirement, the Corps placed a non-competitive order for the classrooms under an existing agreement for portable buildings.

Because the Corps had not been formally assigned this task prior to Katrina's landfall, contracting officials lacked sufficient knowledge of the industry and information about suppliers, inventories, and prices that would have been useful in negotiating a good deal.[61] The business arrangement the Corps used involved layers of subcontractors: the Corps purchased 45 portable buildings from a contractor, who in turn purchased the 45 buildings from a distributor, who in turn purchased them from another distributor, who had purchased the 45 buildings from the manufacturer. Each subcontractor added an additional fee, resulting in the Corps agreeing to a price that was 63 percent higher than the manufacturer's price.

Another example involved the process for ordering and delivering ice. According to Corps officials, FEMA ordered at least double the amount of ice required, resulting in an oversupply of ice and a lack of distribution sites available to handle the volume ordered. One FEMA official working at the local level told us that to ensure that he would receive an adequate amount of ice for first few days, he doubled his initial orders. He said that he kept ordering more ice because headquarters did not provide timely notification of when his shipments would arrive. At the end of hurricane season 2005, FEMA had over 2,000 truckloads of excess ice, which costs the government over $500,000 per month for storage and additional transportation costs. The Senate report on Katrina noted that when Hurricane Katrina hit, and FEMA realized the scope of the temporary housing needs, the agency began buying all of the travel trailers it could find. Unfortunately, FEMA purchased approximately $900 million worth of manufactured homes and modular homes that could not be used because FEMA's own regulations do not allow for these types of homes to be placed in flood plains. Further, some of the homes purchased did not fit FEMA's size standards. However, FEMA seemingly had no plans for how the homes would be used when the purchases were made.

Contract Monitoring and Oversight

Effective monitoring—to ensure that goods and services are delivered in accordance with the agreed upon schedule, cost, quality, and quantity provisions in the contract—relies on having sufficient numbers of trained and properly-deployed personnel to oversee contractor performance. Our work indicated that the number of monitoring staff available was not always sufficient, nor were they effectively deployed to provide sufficient oversight. For example, on FEMA's contracts for installing temporary housing in four states, only 17 of the 27 technical monitors necessary to oversee contractor performance had been assigned at the time of our review. In another case, Corps officials told us that progress in the temporary roof program was slowed due to the lack of sufficient monitors.

Deployment practices did not always provide for appropriate notification of responsibilities or overlap of rotating contracting personnel, thus making knowledge transfer and continuity of contract management operations difficult. For example, for four of the contracts we reviewed, officials were either unaware or not notified by FEMA of their oversight responsibilities. The lack of overlap between oversight personnel for a large temporary housing contract left the most recent contract administrator with no knowledge or documentation of who had authorized the contractor to perform certain activities or why the activities were being performed.

Practices to Help Ensure Successful Acquisition Outcomes

We identified a number of emergency response practices in the public and private sectors that provide insight into how the federal government can better manage its disaster-related procurements. These practices include

- developing knowledge of contractor capabilities and prices by identifying available commodities and services and establishing vendor relationships before they are needed,
- establishing a scalable operations plan to adjust the level of capacity required to effectively respond to the need,
- formally assigning and communicating disaster-related responsibilities, with joint training for government and contractor personnel, and
- providing sufficient numbers of field-level contracting staff with the authority needed to meet mission requirements.

We found a positive example of acquisition practices used during the response to Katrina in our recent review of the Coast Guard's response and recovery missions.[62] Under the NRP, the Coast Guard is the co-lead agency along with the Environmental Protection Agency for Emergency Support Function 10: Oil and Hazardous Materials Response along coastal areas. Prior to Hurricane Katrina, the Coast Guard had basic ordering agreements with fixed prices and other pre-established terms and conditions with private companies to facilitate a rapid response. These agreements were be used to respond to oil spills, for acquiring the use of equipment for vessel salvage and for the use of helicopters. According to Coast Guard officials, they have used basic ordering agreements for this particular mission since the early 1990s, and have extensive knowledge of the market for these particular goods and services. Although we have not conducted a thorough evaluation of the Coast Guard's Marine Environmental Protection clean-up efforts, these agreements contributed to the successful cost control of the marine environmental pollution response after Hurricane Katrina, according to Coast Guard officials.

To help ensure successful acquisition outcomes, we recommend that DHS provide guidance on advance procurement practices and procedures for those federal agencies with roles and responsibilities under the NRP, so that these agencies can better manage disaster-related procurements. These practices should be in advance of disasters, ongoing and continuous, and include (1) developing knowledge of contractor capabilities, and available commodities, services, and prices, as well as developing pre-established vendor relationships, on a competitive basis whenever feasible; (2) establishing scalable operations plans to adjust the level of capacity needed to respond; (3) formally assigning and communicating disaster-related responsibilities and, where feasible, incorporating necessary training; and (4) providing sufficient numbers of field-level contracting staff to meet mission requirements. DHS should also establish an assessment process to monitor agencies' continuous planning efforts for their disaster-related procurement needs and the maintenance of capabilities.

New GAO recommendations to DHS improve advanced procurement practices and procedures
DHS should provide guidance on advance procurement practices and procedures for those federal agencies with roles and responsibilities under the NRP, so that these agencies can better manage disaster-related procurements. These practices should be in advance of disasters, ongoing and continuous, and include (1) developing knowledge of contractor capabilities, and available commodities, services and prices as well as developing pre-established vendor relationships, on a competitive basis whenever feasible; (2) establishing scalable operations plans to adjust the level of capacity needed to respond; (3) formally assigning and communicating disaster-related responsibilities and, where feasible, incorporating necessary training; and (4) providing sufficient numbers of field-level contracting staff to meet mission requirements. DHS should also establish an assessment process to monitor agencies' continuous planning efforts for their disaster related procurement needs and the maintenance of capabilities.
For more detailed information about our recommendations, see appendix I.

Lack of Controls Limited Accountability in the Receipt and Distribution of International Assistance

Federal agencies involved in managing international assistance were not prepared to coordinate, receive, distribute, or account for the assistance. The NRP establishes the Department of State as the coordinator of all offers of international assistance. As part of its Stafford Act responsibilities, FEMA has authority to accept the assistance and coordinate its distribution.[63] Agency officials involved in the cash and in-kind assistance during Hurricane Katrina told us the agencies were not prepared to accept international assistance for use in the United States, because the U.S. government had not received such substantial amounts of international disaster assistance before. Therefore, they told us that they developed ad hoc processes to accept, receive, and distribute the cash and in-kind assistance. Understandably, not all of these ad hoc processes contained controls to support full accountability. For example, we recently reported that no agency tracked and confirmed that the assistance arrived at its destinations.[64] Also, we found that lack of procedures, inadequate information up front about the donations, and insufficient coordination resulted in the U.S. government agreeing to receive food and medical items that were unsuitable for use in the United States and entailed storage costs of about $80,000. DOD's lack of internal guidance regarding the State Department's coordinating process resulted in some foreign military donations that arrived without State Department, FEMA, or DOD oversight.

In the aftermath of Katrina, we recommended that DHS and DOD, in consultation with the Department of State, establish within the NRP—or other appropriate plans—clearly delineated policies and procedures for the acceptance, receipt, and distribution of international assistance to improve the policies, procedures, planning, and oversight of international cash and in-kind donations to the U.S. government in response to disasters. DOD and DHS generally agreed with our recommendations. DHS noted that, in some cases, actions were already underway to address the recommendations.

Recent GAO recommendations to DHS and DOD to improve accountability for international assistance	DOD and DHS responses
DHS and DOD, in consultation with the Department of State, should establish within the NRP—or other appropriate plans—clearly delineated policies and procedures for the acceptance, receipt, and distribution of international assistance.	DOD and DHS generally agreed with our recommendations. DHS noted that, in some cases, actions were already underway.
For more detailed information about our recommendations, see appendix I.	

Control Weaknesses in Individual Assistance Payments Resulted in Fraud and Abuse

When responding to the needs of the victims of a catastrophic disaster, FEMA must balance controls and accountability mechanisms with the immediate need to deliver resources and assistance in an environment where the agency's initial response efforts must focus on life-saving and life-sustaining tasks. Nonetheless, our work has identified flaws in the programs designed to assist disaster victims, which would leave the federal government vulnerable to fraud and abuse of individual assistance payments.

As mentioned earlier, FEMA provides direct assistance—such as temporary housing units—and financial assistance—such as grant funding for temporary housing and other disaster-related expenses—to disaster victims through IHP. Under IHP, FEMA may provide disaster assistance to individuals and households who have uninsured (or underinsured) needs that cannot be met through other means. IHP assistance is limited to 18 months.[65] The maximum amount of financial assistance available is adjusted annually for inflation and was capped at $27,200 in 2006. IHP provides assistance to cover certain expenses not covered by insurance or which individuals or households cannot cover with their own resources. Because IHP benefits are statutorily capped, the program may not cover all losses to damaged property or restore such property to its condition before the disaster.[66] To receive assistance, applicants must qualify for benefits through a process implemented primarily by FEMA contractors and temporary disaster employees in a network of permanent and temporary field offices. The benefits provided to disaster victims of hurricanes in 2005 far surpassed the number of registrants, beneficiaries, and the dollar value amount of benefits provided under the program for named hurricane disasters in 2003 and 2004 combined.

Under the IHP, disaster victims in declared counties must first register and apply for assistance with FEMA, by phone, in person at a disaster recovery center, or over the Internet. FEMA does not provide monetary assistance to any individual without first receiving an application from and taking steps to validate the eligibility of that individual based on the application. For some IHP benefits, applicants above a certain income threshold must first apply to the SBA for disaster loan assistance.[67]

Many of the challenges FEMA faced after Hurricane Katrina stemmed from the magnitude of the disaster, including the number of victims who were displaced from their homes, and related issues in planning, trained staff, and limitations for implementation of the program. As we have reported, FEMA's processes for validating eligibility in the wake of Hurricane Katrina were partially successful. Our work, however, has identified significant flaws in the process for disaster victim applications that leave the federal government vulnerable to fraud and abuse of individual assistance payments.[68]

We estimate that through February 2006, FEMA made about 16 percent, or $1 billion, in improper and potentially fraudulent payments to applicants who used invalid information to apply for disaster assistance. Based on our statistical sample, we are 95-percent confident that the range of improper and potentially fraudulent payments is from $600 million to $1.4 billion. In our assessment of whether a payment was improper and potentially fraudulent, we did not test for other evidence of impropriety or potential fraud, such as insurance fraud and bogus damage claims. This means our review potentially understates the magnitude of improper payments made. Examples of fraud and abuse include payments to applicants who used post office boxes, United Parcel Service stores, and cemeteries as their damaged property addresses. In one case, FEMA paid nearly $6,000 to our applicant who submitted a vacant lot as a damaged address.[69]

For Internet applications, limited automated controls were in place to verify an applicant's identity. However, we found no independent verification of the identity of applicants who applied for disaster assistance over the telephone. To demonstrate the vulnerability inherent in the call-in applications, we used falsified identities, bogus addresses, and fabricated disaster stories to register for IHP and found that we were able to obtain $2,000 expedited assistance checks from FEMA through these means.[70] As we previously testified, FEMA continued to provide our undercover operations with additional disaster-related assistance payments even after FEMA received indications from various sources that our applications may have been bogus.[71] Other control weaknesses included the lack of any validation of damaged property addresses for both Internet and telephone registrations. Our work in this area revealed that thousands of applicants misused social security numbers—i.e., used social security numbers that were never issued or belonged to deceased or other individuals. Our case study investigations of several hundred applications also revealed the use of bogus damaged property addresses. For example, our visits to over 200 of the case study damaged properties in Texas and Louisiana showed that at least 80 of these properties were bogus—including vacant lots and nonexistent apartments. We found that FEMA also made duplicate expedited assistance payments to about 5,000 of the nearly 11,000 debit card recipients—once through the distribution of debit cards and again by check or electronic funds transfer. We found that although debit cards were used predominantly to obtain cash, food, clothing, and personal necessities, a small number were used for adult entertainment, bail bond services, and weapons purchase, which do not appear to be items or services that are essential to satisfy disaster-related needs.

To reduce waste, fraud and abuse in expedited assistance for disaster victims, we have recommended that the Secretary of Homeland Security direct the Undersecretary for Federal Emergency Management to take six actions to address the weaknesses we identified in the administration of IHP: (1) establish an identity verification process for IHP registrants applying via both the Internet and telephone; (2) develop procedures to improve the existing review process of duplicate registrations containing the exact same social security number (SSN); (3) establish an address verification process for IHP registrants applying via both the Internet and telephone; (4) explore entering into an agreement with other agencies to periodically authenticate information contained in IHP registrations; (5) establish procedures to collect duplicate expedited assistance payments or to offset these amounts against future payments; and (6) ensure that any future distribution of IHP debit cards includes instructions on their proper use.[72]

DHS and FEMA concurred fully with four of our six recommendations, and partially concurred with the remaining two recommendations. FEMA and DHS stated that they have already taken actions to address some of these recommendations. These actions include instituting an Internet application process that will prevent all duplicate applications from the Internet, and conducting data sharing tests with the Social Security Administration. In addition, DHS and FEMA stated that, starting in June 2006, all registration addresses (including those provided through phone-in applications) will be subjected to an online verification during the application process. While these are steps in the right direction, we will follow up on whether the actions taken fully address our recommendations. Going forward it will be important for FEMA to establish effective controls to prevent fraudulent and improper payments before they occur, because fraud prevention is a far more effective control than detecting improper and potentially fraudulent payments after they are made. Our experience with organizations that rely on a process that attempts to detect improper and potentially fraudulent payments after they are made is that the organization recovers only a fraction of the payments that should not have been made.

to reduce waste fraud and abuse in the Individuals and Households Program	DHS's responses
Establish an identity verification process for IHP registrants applying via both the Internet and telephone.	FEMA and DHS stated that they have already taken actions to address some of these recommendations, including instituting an Internet application process that will prevent all duplicate registrations from the Internet, and conducting data sharing tests with the Social Security Administration. In addition, DHS and FEMA stated that, starting in June 2006, all registration addresses (including those provided through phone-in applications) will be subjected to an online verification during the registration process. While these are steps in the right direction, we will follow up on whether the actions taken fully address our recommendations.
Develop procedures to improve the existing review process of duplicate registrations containing the exact same SSN.	
Establish an address verification process for IHP registrants applying via both the Internet and telephone.	
Explore entering into an agreement with other agencies to periodically authenticate information contained in IHP registrations.	
Establish procedures to collect duplicate expedited assistance payments or to offset these amounts against future payments.	
Ensure that any future distribution of IHP debit cards includes instructions on their proper use.	
For more detailed information about our recommendations, see appendix I.	

DHS has also reported taking a number of other actions and initiatives designed to improve timeliness and accountability in providing goods and services to the affected areas and their victims. However, because DHS did not provide us documentation to verify these actions and initiatives, we could not determine their status, including the extent to which they are operational. According to DHS, their current efforts are designed to enhance the debris removal guidance, processes and policies to, in part, ensure consistent cost-sharing for federal contracting (through the Corps) and local government contracting. FEMA has also announced a number of customer service improvement efforts so federal recovery programs will have the capacity to handle a catastrophic incident. These include (1) doubling FEMA's registration capacity to 200,000 per day; (2) instituting a pilot project for deployable Mobile Registration Intake Centers; (3) enhancing identity verification during registration; (4) increasing the daily home inspection capacity of FEMA contracted firms from 7,000 per day to 20,000; and (5) updating its policies to improve and quicken determination of applicant eligibility for FEMA's IHP program along with determining eligibility for any expedited assistance available under the program.

Chapter 6

LONG-TERM RECOVERY AND REBUILDING EFFORTS RAISE ISSUES FOR CONGRESS TO CONSIDER

The federal government will be a major partner in the longer-term rebuilding of the Gulf Coast because of the widespread damage and economic impact. Rebuilding raises issues concerning the need for consensus on what rebuilding should be done, where and based on what standards, who will pay for what, and what oversight is needed to ensure federal funds are spent for their intended purposes. In addition, federal programs will face financial difficulties in responding to the long-term needs, and there is uncertainty concerning the impact of catastrophic disasters on the availability and affordability of insurance. Among the issues that will require federal attention include (1) assessing the environmental hazards created by the storms; (2) rebuilding and strengthening the levees; (3) providing assistance to school districts that have enrolled large numbers of evacuee children; (4) continuing to provide assistance for temporary housing, and (5) assuring the financial soundness of the National Flood Insurance Program. Finally, our March 2006 testimony identified guidelines that may enhance federal financial assistance's performance in the restoration of the Gulf Coast.

LONG-TERM RECOVERY IS A SHARED RESPONSIBILITY AND A NUMBER OF FEDERAL AGENCIES WILL PLAY A ROLE IN ADDRESSING MULTIPLE RECOVERY-RELATED ISSUES

State and local officials will have the lead on determining the future needs of the Gulf Coast. However, a number of federal agencies have responsibilities related to the long-term recovery. The recovery should be guided by careful planning that balances the need for speedy economic recovery with actions that reduce the impact of future storms, such as elevating structures located in areas at highest risk of damage from future flooding. In Louisiana and Mississippi, several efforts are underway to implement long-term rebuilding strategies. The actions of state, local, and federal governments, individuals, and nongovernmental entities will affect the speed and nature of the region's recovery.

Our March 2006 testimony identified a number of issues that will require the attention of the Administration and Congress. Among those issues include the rebuilding the region's transportation, health infrastructures and federal facilities, and the availability and affordability of insurance coverage. We also have ongoing work examining other issues that will require federal attention, including (1) assessing the environmental hazards created by the storms; (2) rebuilding and strengthening the levees; (3) providing assistance to school districts that have enrolled large numbers of evacuee children; (4) continuing to provide assistance for temporary housing; and (5) assuring the financial soundness of the National Flood Insurance Program.

Assessing Environmental Hazards Involves the Environmental Protection Agency

Immediately following Katrina, areas along the Gulf Coast faced a number of environmental challenges associated with oil and hazardous material releases resulting from the storm. EPA and federal and state partners continue to monitor air, water, and sediment for potential chemicals of concern such as heavy metals including lead and arsenic; polycyclic aromatic hydrocarbons; pesticides; and diesel and oil range organics and have jointly issued public reports with CDC that provide recommendations on steps individuals can take to limit potential exposure. EPA is also continuing to provide support to Louisiana and Mississippi in assessing drinking water and wastewater infrastructure. Following initial assessments of drinking water and wastewater impacts in Louisiana and Mississippi at the request of the states, FEMA, and the Corps, Louisiana requested additional EPA assistance in conducting needs assessments of repairs at drinking water systems in the state. These facilities are now largely operational, but distribution systems are still being repaired, leaving some areas without service. Through its Office of Recovery and Removal, Mississippi is reviewing wastewater and drinking water needs, and EPA has offered to assist, if needed. While EPA has largely completed its response to hazardous material releases, which has included responding to spills at industrial facilities and collecting orphaned chemical drums and tanks, the agency continues to oversee cleanup of a million-gallon oil spill at a Murphy oil facility in St. Bernard Parish, Louisiana. Finally, EPA continues to assist in hazardous debris removal by coordinating recycling efforts for damaged refrigerators and electronic goods, removing and safely disposing of thousands of household hazardous waste containers such as paint cans and propane tanks, and working with the Army Corps of Engineers and local agencies to remove CFCs and other refrigerants from abandoned appliances that are harmful to the environment.

Areas along the Gulf Coast are also facing environmental challenges as they begin long-term rebuilding efforts. Large-scale demolition efforts now underway along the Gulf Coast create the potential for release of asbestos and other hazardous pollutants. EPA has noted that the number of houses requiring demolition, the sheer volume of debris, and limited landfill space available to accept contaminated debris raise a number of environmental concerns. In addition to EPA's continued role in assisting the Corps and local agencies with the removal of hazardous household waste and appliances, EPA also has a role in ensuring that debris containing asbestos and other pollutants is removed and disposed of appropriately. At the request of the Louisiana and Mississippi Departments of Environmental Quality, EPA has provided some flexibility regarding regulated asbestos materials from homes to facilitate demolition activities in the state of Louisiana and in six counties in Mississippi. This flexibility still requires appropriate practices to ensure protection of public health and the environment. Efforts to reduce the volume of debris by grinding and burning certain types of debris also present environmental challenges, as these activities have the potential to release hazardous air pollutants. EPA continues to work with Louisiana to determine if these practices are appropriate. Continued monitoring will be necessary to minimize the environmental risks associated with demolition and debris removal activities. We are evaluating environmental challenges facing the Gulf Coast, including EPA's oversight of federal asbestos requirements, in an ongoing review of the agency's role in hurricane response.

Rebuilding and Strengthening the Levees Involvesthe U.S. Army Corps of Engineers

We also examined and monitored the Corps plans to repair 169 miles of levees and floodwalls damaged by Hurricane Katrina to pre-storm conditions. Completion of these repairs was planned for June 1, 2006, the start of the 2006 hurricane season. On June 1, 2006, the Corps announced that 100 percent of pre-hurricane levels of protection had been restored although some construction contracts were not yet completed. In instances where the Corps could not complete permanent repairs by June 1, 2006, it made interim repairs and developed emergency procedures to protect against flooding in the event of a hurricane.

After completing these repairs, the Corps planned to (1) repair all pumps, motors and pumping stations by about March 2007; (2) restore sections of existing hurricane protection projects that have settled over time to their original design elevations; and (3) complete construction of incomplete portions of five previously authorized hurricane and flood control projects by September 2007. The Corps also planned to undertake further work to restore, construct, and enhance hurricane protection for southeastern Louisiana by 2010. For example, in April 2006, FEMA released advisory flood elevations for New Orleans and the surrounding area based on a one percent annual chance of flooding, also called a 100-year flood. In response, the Corps is revising its plans and cost estimates to raise the height of levees and floodwalls to provide the area with a 100-year level of protection.

Since September 2005, the Congress has appropriated more than $7 billion to the Corps for portions of this work and additional appropriations are expected. Our ongoing work indicates, however, that the Corps does not have a comprehensive strategy and implementation plan to integrate and manage this work and is currently revising its cost estimates for most system enhancements. Instead, the Corps appears to be following a piecemeal approach, similar to its past practice of building projects without giving sufficient attention to the interrelationships between projects or fully considering whether they will provide an integrated level of hurricane protection for the region. We plan to continue our examination of the Corps' plans and efforts, and to issue a report on this work later this year.

Despite Federal Assistance, States and School Districts Face Continuing Challenges Regarding Funding and Displaced Students

State and local education officials faced challenges in restarting schools and educating displaced students. Hundreds of thousands of students, from kindergarten to the 12th grade, were displaced by the hurricanes. In addition, Louisiana officials said that 29 schools were destroyed and about half of the state's schools were damaged, and Mississippi officials said that 16 schools were destroyed and over half of the state's districts reported some damage. Districts in areas directly affected by the storms and those that enrolled displaced students faced financial challenges. For example, local property tax revenue—a key funding source for schools—may be undercut in areas with property damage, and state funding for schools may also be undercut from the effects of the storms. The large number of displaced students in some districts led to a strain on classroom space, books, teachers, school buses and drivers, and counseling services. Congress initially appropriated a total of approximately $1.4 billion under the Restart and Emergency Impact Aid programs to help reopen elementary and secondary schools and serve displaced students;[73] Congress subsequently appropriated an additional $235 million under the Emergency Impact Aid program for serving displaced students.[74] Federal assistance for displaced students must be obligated by September 30, 2006, and must be used only for expenses incurred during the 2005-2006 school year; yet, state officials reported that a large number of displaced students are likely to remain in their new districts for longer than a year.

In addition, state and district officials expressed the need for flexibility in meeting the requirements of the No Child Left Behind Act of 2001. Upon request of the affected states, the U.S. Department of Education quickly granted some flexibility regarding certain No Child Left Behind Act requirements. State officials were also concerned about the effect of displaced students on their No Child Left Behind Act academic accountability results. As of June 1, 2006, Education had granted several states flexibility with regard to how schools are accountable for the academic achievement of displaced students for the 2005-2006 school year. Although some states received flexibility in how schools are accountable for academic achievement, schools were still responsible for ensuring that displaced students participated in 2005-2006 academic assessments. State and local school officials could promote continuity of operations after large-scale emergencies by developing plans that include locating displaced employees and working closely with other local officials to focus resources on reopening schools. The reopening of schools is vital for community recovery. Also, federal regulatory flexibility in reporting and other requirements can allow states and districts to focus on rebuilding.

Duration of Housing Assistance for Victims of Hurricanes Katrina and Rita Will Require Federal Attention

Our ongoing work on the recovery of the Gulf Coast has identified several issues regarding temporary housing that will require federal attention. These issues generally relate to the question of how long the federal government should provide disaster housing assistance to victims of Hurricanes Katrina and Rita under FEMA and HUD administered programs. Typically, FEMA's IHP provides temporary housing.[75] IHP requires an assessment of eligibility before it can provide housing assistance.[76] According to FEMA officials, in order to house the large number of displaced residents quickly, FEMA used its authority under Section 403 of the Stafford Act to allow states to provide temporary housing for evacuees of Hurricane Katrina. Section 403 of the Stafford Act authorizes the agency to provide assistance essential to meeting immediate threats to life and property resulting from a major disaster, including emergency shelter. By using this authority, FEMA allowed states to house evacuees without assessing eligibility. According to FEMA, approximately 60,000 households received temporary housing under this authority. In early 2006, FEMA began determining the eligibility of those it is assisting under Section 403 to transition to IHP. In March 2006, FEMA announced that it would no longer provide temporary housing benefits under Section 403 as of May 31, 2006.[77] Households deemed ineligible because they do not meet the IHP eligibility criteria will no longer receive housing assistance from FEMA, potentially leaving thousands of people without housing if they are unable to obtain other housing assistance. For example, FEMA officials estimate that approximately 20 percent of the 45,000 households in Texas receiving assistance under Section 403 authority will be ineligible for assistance under IHP.

Victims who receive assistance under FEMA's IHP are eligible to receive temporary housing assistance generally up to 18 months after the date of the disaster declaration. FEMA reported that as of April 2006, it had approved temporary housing assistance for more than 825,000 households displaced by Hurricanes Katrina and Rita. Although some FEMA officials told us that many victims of Hurricanes Katrina and Rita will require housing assistance beyond this limit, it is unclear who will provide it. According to a FEMA official assigned to Louisiana, state and local governments are not currently capable of providing housing assistance after the FEMA assistance ends. FEMA may extend this assistance beyond the 18 month period if it determines that due to extraordinary circumstances an extension would be in the public interest.

For victims of Hurricane Katrina who resided in public or other HUD-assisted housing units damaged or destroyed by the storm, HUD initially provided assistance through its Katrina Disaster Housing Assistance Program. According to HUD, it assisted approximately 15,000 families through this program. In December 2005, Congress appropriated $390 million for temporary rental voucher assistance for victims of Hurricanes Katrina and Rita. [78] Subsequently, HUD established a new program—the Disaster Voucher Program—and in February 2006 began transitioning those from the previous program to the new program. Under the Disaster Voucher Program, households received a housing voucher that covers 100 percent of the rent for up to 18 months. According to HUD guidance, a family is eligible to reoccupy its previously-occupied public or assisted housing unit if and when it becomes available. However, HUD has not yet issued guidance on what housing assistance will be available to displaced residents whose units are not going to be available by the time the assistance ends or those with permanently damaged units[79]. Moreover, as a result of the hurricanes, the housing shortage is even more acute in the affected areas, including public and assisted housing. For example, some residents resided in public housing units that the storms damaged or destroyed. HUD plans to demolish some of its severely damaged public housing stock in New Orleans and other areas affected by the hurricanes.

In response to the hurricanes, HUD's Federal Housing Administration also issued a 90-day moratorium on foreclosures for all FHA-insured loans on properties located in areas affected by Hurricanes Katrina and Rita. HUD subsequently extended the moratorium on foreclosures twice for areas eligible for FEMA's Individual Assistance, because it found that due to magnitude of the storm damage, lenders and borrowers may still need additional time to develop and finalize plans for home repair and resumption of mortgage payments. According to HUD, the last moratorium on foreclosures was extended to August 31, 2006. In addition to the moratoriums, HUD issued guidance instructing lenders servicing FHA-insured mortgage loans not to report hurricane related delinquencies to credit bureaus, not to charge late fees, and to expand their efforts to contact displaced borrowers. Furthermore, HUD offered special mortgage assistance to borrowers in eligible areas who could not maintain mortgage payments due to hurricane related property damage, curtailment of income or increased living expenses. Under this initiative, HUD may make payments to lenders on behalf of borrowers for up to 12 months worth of mortgage payments (principal, interest, taxes, and insurance). This special mortgage assistance is available to eligible borrowers through May 31, 2007. Borrowers are not required to repay HUD until the FHA-insured first mortgage is paid in full. It is unclear, however, what additional assistance lenders and servicers of FHA-insured single-family mortgages may provide and what impact this may have on borrowers and FHA as the guarantor of these mortgages.

We plan to continue our examination of the federal role in providing housing assistance in response to Hurricanes Katrina and Rita, and to issue a report on this work by the end of this year.

Questions Raised about FEMA's National Flood Insurance Program Structure and Long-Term Solvency

Although homeowner insurance policies typically cover damage and losses from fire or theft and often from wind-driven rain, they do not cover flood damage because private insurance companies are largely unwilling to bear the economic risks associated with the potentially catastrophic impact of flooding, including damage from storm surges. To provide some insurance protection for flood victims, as well as incentives for communities to adopt and enforce floodplain management regulations to reduce future flood damage, Congress established the National Flood Insurance Program (NFIP) in 1968.[80] Homeowners with mortgages from federally regulated lenders on property in communities identified to be in special high-risk flood hazard areas are required to purchase flood insurance on their dwellings for, at minimum, the amount of the outstanding mortgage. Optional, lower-cost coverage is also available under the NFIP to protect homes in areas of low to moderate risk. The NFIP provides insurance protection of up to $250,000 for homes and up to $100,000 for personal property.[81] As of December 2005, the NFIP had about 4.8 million policies in force. About 3 million (62 percent) of the policies were for properties in the five states impacted by Hurricanes Katrina and Rita—Alabama, Florida, Louisiana, Mississippi, and Texas.

The NFIP was created in part to reduce taxpayer funded payments to owners of flood-damaged properties by having payments for such damage paid through flood insurance policyholder premiums. The claims from Hurricanes Katrina and Rita required unprecedented borrowing from the Treasury of $18.5 billion at the time of our January 2006 testimony, raising anew questions about its structure and long-term solvency.[82] The program's financial resources are insufficient to meet future expected losses, in part because policy subsidies and repetitive loss properties have contributed to continuing losses to the program. Specifically, the program is not actuarially sound because a high proportion of insurance properties are subsidized—about 26 percent at the time of FEMA's 2004 review. Policy holders for these properties, built before flood plain regulations were established in their communities, pay premiums that represent about 35 to 40 percent of the true risk premium. In

January 2006, FEMA estimated the program had a shortfall of $750 million in annual premium income because of policy subsidies.

The portion of subsidized polices that most adversely impact the program's financial solvency are about 49,000 repetitive loss properties (as of March 2004) for which two or more claims of $1,000 or more have been paid in a 10-year period. Although these properties make up only about 1 percent of the properties insured under the NFIP, they account for 25 to 30 percent of all claims losses. As of March 2004, nearly half of all nationwide repetitive loss property insurance payments had been made in Louisiana, Texas, and Florida. These properties accounted for about $4.6 billion in claims payments from 1978 to March 2004. A significant number of repetitive loss properties were affected by Hurricanes Katrina and Rita, and a FEMA management official predicted that the inventory of repetitive loss properties would increase as a result of damage from the two storms.

As part of its flood plain management strategy, NFIP policies encourage states and local communities to elevate or remove damaged properties from the flood plain. In addition to paying claims for flood damage, NFIP policies pay up to $30,000 for the cost of complying with mitigation actions required under state or local floodplain management laws or ordinances, such as elevating, moving, or demolishing the damaged structures. This increased cost of compliance (ICC) coverage is available under the NFIP's standard flood insurance policy for properties that suffer substantial damage, including repetitive loss properties.[83] In an upcoming revision to the standard flood insurance policy, FEMA plans to make permanent the increase in time to complete work and receive an ICC payment. It is too early in the recovery process to determine the impact of the mitigation actions on the national inventory of repetitive loss properties.

For all these reasons, and others described in our March 2006 testimony, we have placed the NFIP on our list of high-risk government programs. In 2004 we made recommendations to FEMA on strategies to better ensure that FEMA's map modernization achieves the intended benefits of improved flood mitigation, increased flood insurance participation, and improved multi-hazard mitigation and risk management capabilities.[84] DHS and FEMA generally agreed with our recommendations. FEMA said that it planned to refine existing standards, in coordination with stakeholders, to ensure consistent data collection and analysis for all communities commensurate with their flood risk; that it would continue to collaborate with stakeholder groups to develop an effective strategy to include states and communities with varying levels of capabilities and resources; and that it planned to refine performance measures for this map modernization objective to make them more useful and quantifiable. In 2005, we recommended that FEMA use a statistically valid method to select claims for quality review because its existing sampling and internal controls did not provide management with the information needed to have reasonable assurance that program objectives are being achieved. We also recommended that FEMA set target dates for implementing the provisions of the Flood Insurance Reform Act of 2004.[85] FEMA said its existing sampling method was sufficient for the purposes for which it was used and that it was working diligently to implement the requirements of the 2004 reform act.[86]

Recent GAO recommendations to DHS to enhance oversight and management of NFIP	DHS responses
Flood Map Modernization 2004 FEMA should ensure that its map modernization achieves the intended benefits of improved flood mitigation, increased flood insurance participation, and improved multi-hazard mitigation and risk management capabilities.	FEMA said that it planned to refine existing standards, in coordination with stakeholders, to ensure consistent data collection and analysis for all communities commensurate with their flood risk; that it would continue to collaborate with stakeholder groups to develop an effective strategy to include states and communities with varying levels of capabilities and resources; and that it planned to refine performance measures for this map modernization objective to make them more useful and quantifiable.
National Flood Insurance Program 2005 FEMA should use a statistically valid method to select claims for quality review because existing sampling and internal controls did not provide management with the information needed to have reasonable assurance that program objectives are being achieved.	FEMA said its existing sampling method was sufficient for the purposes for which it was used.
FEMA should set target dates for implementing the provisions of the Flood Insurance Reform Act of 2004.	FEMA responded that it was working diligently to implement the requirements of the 2004 reform.
For more detailed information about our recommendations, see appendix I.	

A Framework to Enhance Federal Financial Assistance's Performance in the Restoration of the Gulf Coast

Finally, in our March 8 testimony, we identified an accountability and performance framework from relevant GAO past work that may be considered regarding the provision of federal financial assistance for the restoration of the Gulf Coast:[87]

- Identify the scope of the problem. For example, does the problem reflect broader industry wide or regional economic conditions? For the Gulf Coast, this would involve financial and economic analyses, perhaps utilizing current studies of prior conditions and the ongoing progress of recovery and rebuilding.
- Clearly establish the effect of the problem on the national interest—such as whether federal involvement is needed because the problem presents potentially large economy wide or regional consequences. For example, in the Gulf Coast, Congress should consider whether the proposed rebuilding plans are reasonable and the involvement of state and local governments and the private sector will not, on their own, provide necessary capital.
- Establish clear, concise, and consistent legislative goals and objectives associated with the response. For example, in the Gulf Coast, building on the President's decision to appoint a Coordinator of Federal Support for the Recovery and Rebuilding of the Gulf Coast region, Congress should ensure that the goals of all aspects of federal involvement are clear, measurable, and agreed-upon by all participants.

- Protect the government's financial interest. In the Gulf Coast, for example, controls might be put in place so there is review of the most important financial and operating plans that rely on federal investments.

In terms of protecting the government's financial interest, our ongoing work in this area found that there is no one agency or central collection point that exists to compile and report on how the emergency supplemental appropriations funds provided to 23 different federal agencies are being spent. Without a framework and mechanisms in place to collect and consolidate information from these agencies on a periodic basis, it will be difficult for decision makers to determine how much federal funding has been spent and by whom, whether more may be needed, or whether too much has been provided. The ability to separately track and report on these funds is important to help ensure better accountability and clearly identify the status of funding provided in direct response to these hurricanes at both the individual federal agency level as well as the government wide level. Also, it is important to provide transparency so that hurricane victims, affected states, as well as American taxpayers, know how these funds are being spent. We will issue a report later this year that addresses the federal government's ability to track and report on the hurricane relief funds received.

Chapter 7

CONCLUSIONS

Since September 11, 2001, the federal government has awarded billions of dollars in grants and assistance to state and local governments to assist in strengthening emergency management capabilities. DHS has developed several key policy documents, including the NRP, NIMS, and the interim National Preparedness Goal to guide federal, state, and local efforts. The aftermath of the 2005 hurricane season resulted in a reassessment of the federal role in preparing for and responding to catastrophic events. The studies and reports of the past year—by Congress, the White House Homeland Security Council, the DHS-IG, DHS and FEMA, GAO, and others—have provided a number of insights into the strengths and limitations of the nation's capacity to respond to catastrophic disasters and resulted in a number of recommendations for strengthening that capacity. Collectively, these studies and reports paint a complex mosaic of the challenges that the nation—federal, state, local, and tribal governments, nongovernmental entities, the private sector, and individual citizens—faces in preparing for, responding to, and recovering from catastrophic disasters. In addition to the reports issued to date, there are numerous ongoing analyses of various facets of the nation's preparedness and response efforts before and after Hurricane Katrina.

Improving the nation's ability to prepare, respond, and recover from catastrophic disasters will take three critical inputs: (1) leaders and professionals with the right knowledge, skills, and experience; (2) plans and guidance that detail what needs to be done, by whom, how, and how well; and finally (3) clear criteria and expectations that are clearly communicated, well understood, and result in appropriate, coordinated actions from all levels of government, their emergency planners and responders, and the nonprofit and private-sector organizations that will be providing support. This requires the development of thoughtful strategic planning and assessment, along with developing and sustaining needed skills and assets. Effective training and exercises based on realistic scenarios is a key component of building and maintaining needed skills and capabilities. As we noted in past GAO work, overall federal assistance has not been guided by a clear, risk-based strategic plan that would provide a basis for realistic budgeting and resource planning.[88] Ultimately, the federal government must determine how much it will cost to develop and maintain these needed capabilities and what the federal government can afford to pay. Other levels of government face a similar challenge.

DHS and its partners and stakeholders—governmental and nongovernmental, public and private—face the challenge of working together to coordinate preparedness activities and formulate realistic budgets and resource plans to share these costs and support and sustain implementation of an efficient and effective all-hazards national preparedness program.

Catastrophic disasters are unique in their scope and the magnitude of their effects. In preparing for any disaster, but particularly catastrophic disasters, it is essential to have in place through contracts, prepositioning of key supplies, and other means the needed surge capacity to respond quickly and effectively to the destruction and dislocation that results from the catastrophe. In moving forward, one critical challenge will be determining if the initial and long-term efforts to implement specific initiatives will truly close the identified gaps in the nation's capacity to prepare for, respond to, and recover from catastrophic disasters. Sound recommendations and initiatives must be effectively implemented to achieve the intended improvements. Currently, there is little available information on the operational readiness of many of the reforms and actions DHS has announced in recent months. The first real test of these actions will come with the next major hurricane or other major disaster. DHS has conducted some exercises in recent weeks, but there is little available information on the results of those exercises. In addition, some of the reforms, such as revisions to the NRP, may have clarified some issues, such as the role of the Secretary of Homeland Security in declaring incidents of national significance, while potentially raising new issues, such as how the NRP, which DHS now states is in effect at all times, would be operationalized in incidents of lesser severity. While the scope of the NRP has broadened, DHS and other federal agencies may be without a road map for dealing with incidents of lesser severity until they supplement the NRP with detailed, scalable operational plans. Likewise, DHS and other federal agencies may be without a road map in responding to catastrophic incidents until they supplement the NRP's catastrophic incident annex with the detailed operational plans envisioned by the NRP. To be effective, the NRP must be supported by robust operational plans for implementing its provisions.

Appropriate controls and accountability mechanisms for the use of resources during a catastrophic disaster are essential to ensure that the resources are used appropriately, but there is always a tension between normal controls and accountability mechanisms and the need to deliver assistance expeditiously, especially when responders and governments are providing life-saving and life-sustaining services in a time-critical environment. DHS and state and local governments all face a significant challenge in ensuring that relief payments and services are only sent to valid registrants while also distributing those relief payments and services as fast as possible. Thus, all levels of governments must further develop and strengthen controls to ensure accountability, because—as FEMA has learned from prior experience—pursuing collection activities after disaster relief payments have been made is costly, time-consuming, and ineffective. Upfront controls are all the more crucial given the estimated billions of dollars in erroneous or excessive payments related to Hurricanes Katrina and Rita.

Finally, given the magnitude of Hurricane Katrina's devastation of the physical and economic infrastructure in the Gulf Coast—both public and private sector—rebuilding is likely to take years, if not decades, to complete. As a result, all levels of government will have a critical role in the effort, not just the federal government and not solely or even primarily through FEMA's post-disaster recovery and mitigation funding. Some communities were so totally devastated that they almost face rebuilding their communities from the ground up. These long-term recovery and rebuilding efforts offer an opportunity to mitigate the potential impact of future hurricanes, and employ both direct governmental funding and forms of fiscal and monetary support from the banking and insurance industries. In light of how long this effort may take, GAO will continue to examine current recovery issues and long-term rebuilding activities.

Chapter 8

RECOMMENDATIONS

In this report we are making several new recommendations updating and formalizing several recommendations first posed in our March 8, 2006, testimony on preliminary observations regarding preparedness, response, and recovery; and one new recommendation on advance procurement practices and procedures:

RECOMMENDATIONS FOR EXECUTIVE ACTION

- Rigorously re-test, train, and exercise its recent clarification of the roles, responsibilities, and lines of authority for all levels of leadership, implementing changes needed to remedy identified coordination problems.
- Direct that the NRP base plan and its Catastrophic Incident Annex be supported by more robust and detailed operational implementation plans, particularly the Catastrophic Incident Supplement to the NRP. Such operational plans should, for example, further define and leverage those military capabilities that might be needed in a catastrophic disaster.
- Provide guidance and direction for federal, state, and local planning, training, and exercises to ensure such activities fully support preparedness, response, and recovery responsibilities at a jurisdictional and regional basis. This should also include the application of lessons learned from actual catastrophic and other disasters.
- Take the lead in monitoring federal agencies' efforts to meet their responsibilities under the NRP and the interim National Preparedness Goal, including the development, testing, and exercising of agency operational plans to implement their responsibilities under the NRP, NIMS, and the interim National Preparedness Goal.
- Given that resources are finite, apply an all-hazards, risk management approach in deciding whether and how to invest in specific capabilities for a catastrophic disaster.
- Provide guidance on advance procurement practices and procedures for those federal agencies with roles and responsibilities under the NRP, so that these agencies can better manage disaster-related procurements, such as food, shelter, and debris removal. These practices should be in advance of disasters, ongoing and continuous, and include (1) developing knowledge of contractor capabilities and available commodities, services and prices as well as developing preestablished vendor

relationships, on a competitive basis whenever feasible; (2) establishing scalable operations plans to adjust the level of capacity needed to respond; (3) formally assigning and communicating disaster-related responsibilities and, where feasible, incorporating necessary training; and (4) providing sufficient numbers of field-level contracting staff to meet mission requirements. DHS should also establish an assessment process to monitor agencies' continuous planning efforts for their disaster-related procurement needs and the maintenance of capabilities.

MATTERS FOR CONGRESSIONAL CONSIDERATION

Reaffirming a recommendation we made following Hurricane Andrew, we recommend that Congress:

- Give federal agencies explicit authority to take actions to prepare for catastrophic disasters when there is warning. We also offer some analytical frameworks and factors that Congress may wish to consider in carrying out its oversight and legislative responsibilities with regard to national preparedness and the recovery of the Gulf Coast region:
- Use a risk management framework to assist in its oversight and legislative decision-making regarding the nation's capacity to respond to catastrophic disasters.
- If Congress is considering a change in FEMA's organizational placement, it should consider (1) whether factors such as the qualifications, experience, and training of the leadership and the adequacy of resources led to its performance difficulties; (2) criteria such as mission relevancy, similar goals, and objectives (present and future); (3) leveraging the effectiveness of other agencies and programs or the new department as a whole; and (4) gains in efficiency and effectiveness through eliminating duplications and overlaps.
- If Congress is considering a change in the qualifications of the Undersecretary for Federal Emergency Management, it should consider establishing statutory professional qualifications for the Undersecretary and other selected key positions within DHS and term appointments for the Undersecretary and selected other positions.
- Consider the four conditions that we suggested in 1984, as a framework of ideas about how to structure future financial assistance programs and what program requirements to include to achieve Congressional goals and objectives while minimizing the risk of financial loss to the government. These guidelines are a useful framework for developing assistance programs for the Gulf Coast restoration:
- The scope of the problem should be identified, such as if the problem reflects broader industry wide or regional economic conditions. For the Gulf Coast, this would involve financial and economic analyses, perhaps utilizing current studies of prior conditions and the ongoing progress of recovery and rebuilding.
- The effect of the problem on the national interest should be clearly established, for example, whether the problem presents potentially large economy wide or regional consequences. For example, in the Gulf Coast, Congress should consider the costs of

municipal and corporate collapse and the challenges associated with providing assistance.

- The legislative goals and objectives associated with the response should be clear, concise, and consistent. For example, in the Gulf Coast, goals and objectives for rebuilding should be clearly stated, working with the state and local groups already tasked with recovery planning and with the Administration's Coordinator of Federal Support for the Recovery and Rebuilding of the Gulf Coast region.
- Lastly, the government's financial interest should be protected. In the Gulf Coast, controls might be put in place so that the most important financial and operating plans will be reviewed.

This report also identifies the major findings, conclusions, and 74 recommendations and matters for congressional consideration from our prior and ongoing work on catastrophic disasters. Appendix I summarizes 24 key GAO recommendations identified in work prior to Hurricane Katrina. These 24 prior recommendations were not adopted or in effect when Hurricane Katrina hit the Gulf Coast and remain listed by GAO as open recommendations, that is, recommendations not fully implemented. We continue to believe that, for the most part, these recommendations are still viable. In addition, appendix I lists 43 GAO recommendations from GAO reports in the aftermath of Hurricanes Katrina and Rita, and 7 new recommendations formalized in this report.

AGENCY COMMENTS

We provided a draft of this report to DHS for review and comment, and also provided relevant sections of the draft report to various federal departments and agencies including the departments of Agriculture, Education, HHS, HUD, Labor, and State, as well as SBA, EPA and Social Security Administration. The Departments of State and HHS said they had no comments on our draft. The Department of Agriculture's Food and Nutrition Service, the Department of Education, EPA, HUD, and the Social Security Administration provided additional technical comments that we incorporated, and the Social Security Administration also stated that they were pleased to be included in the report as an example of good planning.

DHS provided written comments on August 28, 2006, signed by the Undersecretary for Federal Emergency Management and the Undersecretary for Preparedness. DHS's comments are reproduced in full in appendix II. DHS also provided technical comments that we incorporated as appropriate. DHS generally concurred with the six new recommendations in our draft report and described an array of actions it has taken, has underway, or planned to implement those recommendations. DHS also described actions it has taken to address the expansion of search and rescue capabilities; the supply pre-positioning and tracking of emergency supplies, such as food, ice, and water; FEMA staffing; and the acceptance and management of international donations for disaster response and relief.

If effectively implemented, the actions that DHS described should basically address the problems that we described in the draft report. However, as we noted in our report, the NRP revisions may not fully resolve the leadership issues with respect to the PFO and FCO roles and questions remain with regard to how the NRP, which now states it is in effect at all times would be operationalized in incidents of severity less than the incidents of national significance that are to be managed by the Secretary of Homeland Security. Because we did not have time to evaluate DHS' actions completed to date, we cannot reach any conclusions regarding the extent to which those actions are fully operational and have improved disaster preparedness and response capabilities.

In commenting on our recommendations, DHS stated that planning for patient evacuation out of hospitals and nursing homes is being coordinated by HHS, as Coordinating Agency for the NRP's Emergency Support Function 8. We understand that responding to our recent recommendation of planning for patient evacuation out of hospitals and nursing homes will involve coordination among various agencies, including HHS, and addressed our recommendations to DHS because it is responsible for the NRP, the Catastrophic Incident Annex and its Supplement, and the activation of NDMS.[89]

On August 14, 2006, we received written comments from the Department of Labor who stated that our report correctly points out that OSHA and FEMA had experienced difficulties in agreeing on roles and procedures. OSHA has reported that their agency and FEMA have worked together to develop procedures for role of the Safety and Health Coordinator in the Joint Field Office and for the NRP's Worker Safety and Health Support Annex.

SBA's Associate Administrator for Disaster Assistance provided written comments on August 14, 2006, on a draft of the SBA segment of this report. SBA's comments basically reiterate the comments it made on our earlier report [90] and are reproduced in full in appendix III. SBA stated that more emphasis should have been given to its improvement efforts and the benefits of DCMS compared with its previous system and stated its concerns regarding the use of catastrophe risk model data in determining the user requirements of DCMS. It was not within the scope of our work to conduct a comparative analysis of DCMS and SBA's previous system, but we recognized some of the benefits the agency realized by adopting DCMS. We continue to believe that catastrophe risk modeling firms provide critical information, such as the likelihood and severity of damages from potential catastrophes. Combined with other elements of a comprehensive planning process, such information would have been useful in planning the maximum user capacity of DCMS. If SBA had considered this information, it may have expanded the maximum user requirement for DCMS and been better prepared to reduce the backlog of loan applications.

APPENDIX I: SUMMARY OF KEY OPEN GAO RECOMMENDATIONS ON CATASTROPHIC DISASTERS

The following three tables show GAO's recommendations on dealing with catastrophic disasters made before Hurricane Katrina, in the aftermath of Katrina, and new recommendations formalized in this report. The tables also show the agency response to each recommendation. Recommendations are given in chronological order in each table, with the most recent recommendations listed first.

Table 4. Key Open Recommendations Made Prior to Hurricanes Katrina and Rita[a]

Title and GAO product number	Summary of recommendations	Agency response
Reserve Forces: Actions Needed to Better Prepare the National Guard for Future Overseas and Domestic Missions, GAO-05-21	Recommendations to the Secretary of Defense: To improve the Army and the Air National Guard's preparedness to perform homeland security missions, the Secretary of Defense should: • Assess how support for current operations will affect the readiness of non-deployed Army National Guard forces for future overseas and domestic missions. • Specify how the Army plans to restructure and provide the Guard resources—personnel, equipment, and training—consistent with its 21st century role. • Establish the full range of the National Guard's homeland missions, including those led by DOD and those conducted in support of civilian authorities. • Identify the National Guard's capabilities to perform these missions and any shortfalls in personnel, equipment, and training needed to perform these missions successfully. • Develop a plan to manage the risk associated with the declining readiness of non-deployed Army National Guard forces, including identifying funding for any personnel and equipment required to mitigate unacceptable levels of risk. • Establish readiness standards and measures for the Guard's homeland security missions so that the readiness for these missions can be systematically measured and accurately reported.	The department generally agreed with our recommendations and cited actions it was taking to implement them. DOD partially agreed with our recommendation that DOD develop plans to manage the risk associated with the declining readiness of non-deployed Army National Guard forces. We agreed that the Army Campaign Plan is a significant step in planning to address National Guard readiness problems because it identifies goals and objectives and assigns responsibilities for actions to plan for transforming its forces. However, we believed the Army Campaign Plan did not fully meet the intent of our recommendation because it lacked specificity about how the Army would address the readiness of non-deployed Army National Guard forces in the near term and how all Guard units will be converted to the modular design. DOD concurred with our recommendation to establish the full range of the National Guard's homeland missions, to identify the capabilities needed to perform those missions and develop a plan to address any shortfalls, and to establish readiness standards and measures for the Guard's homeland security missions. However, in its comments, DOD said it would take a different approach to accomplishing the tasks than we recommended. We believe the approach DOD proposes meets the intent of our recommendation, and we have modified the wording of our recommendation to reflect the proposed change in organizational responsibilities.

Table 4. (Continued).

Title and GAO product number	Summary of recommendations	Agency response
Leadership and Intergovernmental Cooperation Required to Achieve First Responder Interoperable Communications, GAO-04-740	Recommendations to the Secretary of Homeland Security: To address a fundamental barrier to successfully addressing the decades-old problems of interoperable communications—the lack of effective, collaborative, interdisciplinary, and intergovernmental planning, the Secretary of Homeland Security should: • Work with the Federal Communications Commission and the National Telecommunications and Information Agency to develop a nationwide database of interoperable communications frequencies and a common nomenclature for those frequencies and establish clear timeframes to complete both efforts. • Establish interoperability requirements whose achievement can be measured. • Through grant guidance, encourage states to establish a single statewide body that will develop a comprehensive statewide interoperable communications plan. • Require that federal grant funding of new equipment shall be approved only upon certification by the statewide body that such grants conform with the statewide interoperability plan. • In conjunction with the Director of OMB, review the interoperability mission and functions now performed by Wireless Public Safety Interoperable Communications Program and establish these functions as a long-term program with adequate coordination authority and funding.	DHS said that it was developing a nationwide database of interoperable public safety communications frequencies and also said it planned to work on a common nomenclature across public safety disciplines and jurisdictions. With respect to our second recommendation, DHS said it would develop a methodology to establish a national baseline of public safety communication and interoperability capabilities with input from the public safety community. However, the DHS letter did not directly address our recommendation about encouraging states to create statewide bodies for interoperable communications that would establish statewide interoperability plans for federal, state, and local communications systems in all frequency bands.

Table 4. (Continued).

Title and GAO product number	Summary of recommendations	Agency response
Homeland Security: Management of First Responder Grants in the National Capital Region Reflects the Need for Coordinated Planning and Performance Goals, GAO-04-433	Recommendations to the Secretary of Homeland Security: To help ensure that emergency preparedness grants and associated funds are managed in a way that maximizes their effectiveness, the secretary should: • Work with the National Capital Region jurisdictions to develop a coordinated strategic plan to establish goals and priorities for enhancing first responder capacities that can be used to guide the use of federal emergency preparedness funds. • Monitor the plan's implementation to ensure that funds are used in a way that promotes effective expenditures that are not unnecessarily duplicative. • Identify and address gaps in emergency preparedness and evaluate the effectiveness of expenditures in meeting those needs by adapting standards and preparedness guidelines.	DHS generally agreed with our recommendations but also stated that National Capital Region jurisdictions had worked cooperatively together to identify opportunities for synergies and lay a foundation for meeting the challenges noted in the report. DHS agreed that there is a need to continue to improve preparedness by developing more specific and improved preparedness standards, clearer performance goals, and an improved method for tracking regional initiatives. However, DHS and GAO disagreed about oversight roles and responsibilities.
Flood Map Modernization: Program Strategy Shows Promise, but Challenges Remain, GAO-04-417	Recommendations to the Secretary of Homeland Security: To help ensure that FEMA's map modernization achieves the intended benefits of improved flood mitigation, increased flood insurance participation, and improved multi-hazard mitigation and risk management capabilities through the production of more accurate and accessible flood maps the Secretary of Homeland Security should direct the Undersecretary of Emergency Preparedness and Response to take the following four actions: • Develop and implement data standards that will enable FEMA, its contractor, and its state and local partners to identify and use consistent data collection and analysis methods for communities with similar risk. • Develop and implement strategies for partnering with state and local entities with varying levels of capabilities and resources. • Ensure that it has the staff capacity to effectively implement the nationwide mapping contract and the overall map modernization program. • Develop and implement useful performance measures that define FEMA' s progress in increasing stakeholders' awareness and use of the new maps, including improved mitigation efforts and increased participation rates in purchasing flood insurance.	DHS and FEMA generally agreed with our recommendations. FEMA said that it planned to refine existing standards, in coordination with stakeholders, to ensure consistent data collection and analysis for all communities commensurate with their flood risk; that it would continue to collaborate with stakeholder groups to develop an effective strategy to include states and communities with varying levels of capabilities and resources; that it had begun planning to ensure that additional staffing needs are met; and that it planned to refine performance measures for this map modernization objective to make them more useful and quantifiable.

Table 4. (Continued).

Title and GAO product number	Summary of recommendations	Agency response
Continuity of Operations: Improved Planning Needed to Ensure Delivery of Essential Government Services, GAO-04-160	Recommendations to the Secretary of Homeland Security: To enhance the ability of the executive branch to continue to provide essential services during emergencies, the secretary should: • Direct the Undersecretary of Preparedness to take steps to improve the oversight of continuity of operations planning by ensuring that agencies correct the deficiencies in individual continuity of operations plans identified here, as well as those identified in previous assessments. • Conduct assessments of agency continuity plans that include independent verification of agency-provided information, as well as an assessment of the essential functions identified and their interdependencies with other activities.	DHS agreed that better continuity of operations planning is needed to ensure delivery of essential services and that FEMA could do more to improve continuity of operations planning and has begun to correct the identified deficiencies. DHS has since reported to GAO that it had taken several actions to improve agency continuity of operations plans, including conducting an interagency exercise in 2004 and providing training to more than 350 continuity of operations planning managers at 65 agencies. In addition, DHS officials reported that after it fully deploys a Readiness Reporting System to track agency compliance with FEMA guidance, agencies will be required to periodically report to FEMA on their compliance. DHS reports that its Readiness Reporting System will be used to assist it in assessing agency contingency plans using information provided by the agencies. Agencies will then be required to provide periodic updates on their level of compliance with FEMA guidance.

Table 4. (Continued).

Title and GAO product number	Summary of recommendations	Agency response
Disaster Management: Recent Disasters Demonstrate the Need to Improve the Nation's Response Strategy, GAO/T-RCED-93-46; Disaster Management: Improving the Nation's Response to Catastrophic Disasters, GAO/RCED-93-186; and Disaster Assistance: DOD's Support for Hurricanes Andrew and Iniki and Typhoon Omar, NSIAD-93-180	In 1993, in response to Hurricane Andrew, we conducted several reviews and made recommendations regarding the unique challenges involved in responding to catastrophic disasters. Recommendations to FEMA: • FEMA should improve its catastrophic disaster response capability by using existing authority to aggressively respond to catastrophic disasters, assessing the extent of damage, and then actively advising state and local officials of identified needs and the federal resources available to address them, as well as the extent to which DOD resources will be needed to supplement those of the Red Cross in meeting mass care needs. • FEMA should enhance the capacity of state and local governments to respond to catastrophic disasters by (a) continuing to give them increasing flexibility to match grant funding with individual response needs, (b) upgrading training and exercises for catastrophic disaster response, and (c) assessing each state's preparedness for catastrophic disaster response. Matters for Congressional Consideration: Congress should consider • Giving FEMA and other federal agencies explicit authority to take actions to prepare for catastrophic disasters when there is warning. • Removing statutory restrictions on DOD's authority to activate Reserve units for catastrophic disaster relief.	FEMA generally agreed with our findings and recommendations and had begun taking actions to address them. These included a proposal in FEMA's fiscal year 1994 budget request to restructure in a manner it felt came close to our recommendation that FEMA form a disaster unit. In addition, FEMA told us it would revise and update its training and exercises on the basis of the lessons learned from recent catastrophic disasters; developing methods to better determine state and local preparedness for disaster response; and seeking clarified legislative authority when a catastrophic disaster is predicted. At the time we said FEMA's recent initiatives are a good start toward improving its response to future catastrophic disasters. However, we noted that some of the improvements necessary in the overall federal response are outside of FEMA's control. Presidential action is needed to strengthen leadership for the federal response by designating a key White House official to oversee preparedness and response activities. Furthermore, legislative action is needed to improve all federal agencies' preparedness when there is warning of a disaster and to allow DOD to activate Reserve units for catastrophic disaster relief. However, FEMA's mission, placement, and all-hazards approach have been re-configured several times and the recommendations to FEMA have remained open. We did subsequently closed our matters for congressional consideration when it appeared Congress would not take any action on the removing restrictions on Reserve activations, and when it considered legislation for giving agencies explicit authority to prepare for pending catastrophic disaster. However, given the failures witnessed in Hurricane Katrina and the new threat of terrorism we are reopening these recommendations to Congress.

Table 5. Recent Open Recommendations Made in the Aftermath of Hurricanes Katrina and Rita[a]

Title and GAO product number	Summary of recommendations	Agency response
Federal Action Needed to Ensure States Have Plans to Safeguard Children in the Child Welfare System Displaced by Disasters, GAO-06-944	Recommendations to the Secretary of Health and Human Services: To better assist states in developing child welfare disaster plans, we are recommending that the Secretary of Health and Human Services • Ensure that the department's child welfare disaster planning guidance address the dispersion of children and families within and across state lines. This guidance should include information on • preserving child welfare records, identifying children who may be dispersed, • identifying new child welfare cases and providing services, • coordinating services and sharing information with other states, and placing children from other states. • Develop and provide training on child welfare disaster planning to all states. Recommendations to Congress: To ensure continuity of services within or across state lines for the children under state care, Congress should consider requiring that states develop and submit child welfare disaster plans for HHS review.	Regarding our recommendations that HHS ensure that the department's guidance and training to states on child welfare disaster planning address the dispersion of children and families, the Administration for Children and Families stated that it has taken action to update the guidance and provide training to states and will encourage them to develop and submit disaster plans for review.
SMALL BUSINESS ADMINISTRATION: ACTIONS NEEDED TO PROVIDE MORE TIMELY DISASTER ASSISTANCE, GAO-06-860	In order to provide more timely disaster assistance in the future, we recommend that the Administrator of SBA direct its Office of Disaster Assistance to take the following four actions: • Reassess the Disaster Credit Management System (DCMS) maximum user capacity and related loan processing resource needs based on such things as lessons learned from the Gulf Coast hurricanes, a review of information available from catastrophe risk modeling firms and disaster simulations, and related cost considerations. • Conduct complete stress testing to ensure that DCMS can function at planned for maximum user capacity levels. • Improve management controls over assessing contractor performance through inspections of all equipment purchased or leased to support DCMS. • Expedite plans to resume business processing reengineering efforts to analyze the disaster loan process and identify ways to more efficiently process loan applications, including an evaluation of the feasibility of implementing a secure Internet-based application feature for home loan applicants.	SBA stated that it generally agreed with our recommendations and intended to improve the delivery of the Disaster Loan Program for events of all sizes. However, SBA disagreed with some of the findings and conclusions in our draft report.

Table 5. (Continued).

Title and GAO product number	Summary of recommendations	Agency response
Disaster Evacuations: Limitation in Federal Assistance to Health Facilities for Transportation Should be Addressed, GAO-06-826	Recommendations to the Secretary of Homeland Security: To address limitations in how the federal government provides assistance with the evacuation of health care facilities, we recommend that the Secretary of Homeland Security take the following two actions: • Clearly delineate how the federal government will assist state and local governments with the movement of patients' and residents out of hospitals and nursing homes to a mobilization center where National Disaster Medical System (NDMS) transportation begins. • In consultation with the other NDMS federal partners—the Secretaries of Defense, Health and Human Services, and Veterans Affairs—clearly delineate how to address the needs of nursing home residents during evacuations, including the arrangements necessary to relocate these residents.	DHS stated that it will take our recommendations under advisement as it reviews the National Response Plan. According to DHS, all of the NDMS federal partners are currently reviewing the NDMS memorandum of agreement with a view towards working with state and local partners to alter, delineate, and otherwise clarify roles and responsibilities as appropriate. However, we noted that, as stated in the draft report, neither NDMS documents, the NRP, nor the draft Catastrophic Incident Supplement to the NRP—to be used in cases when the capabilities of state and local governments are almost immediately overwhelmed—describe the federal role in coordinating with state and local authorities during hospital and nursing home evacuations. We also noted that reliance on state and local resources was inadequate when multiple facilities in a community had to evacuate simultaneously. DOD disagreed with our conclusions concerning the federal role in short distance transportation and the successful evacuation of nursing home residents during Hurricane Rita. However, during a catastrophic incident, the capabilities of state and local governments may almost immediately become overwhelmed. As we stated above in our response to DHS's comments, the federal role in these situations has not been described. Second, Our draft report did describe NDMS's evacuation of people, including nursing home residents. However, we also noted that the NDMS after-action report on hurricanes Katrina and Rita states that NDMS was not optimally prepared to manage the nursing home requirements of evacuees who did not require hospitalization. For this reason, we believe that explicit consideration of the needs of nursing home residents is warranted. HHS and Veterans Affairs concurred with our recommendations.

Table 5. (Continued).

Title and GAO product number	Summary of recommendations	Agency response
Continuity of Operations: Selected Agencies Could Improve Planning for Use of Alternate Facilities and Telework during Disruptions, GAO-06-713	Recommendations to the Secretary of Homeland Security: To improve the assessment and oversight of agency continuity planning and developing guidance on including telework in such planning, and to ensure that agencies are adequately prepared to continue performing essential functions following an emergency, the Secretary of Homeland Security should direct the Undersecretary for Federal Emergency Management to: • Before the upcoming interagency exercise, conduct an assessment of the continuity of operations plans of the six agencies whose alternate facilities were included in our review and report any deficiencies to the head of the agency for correction. • Develop a methodology for individual agency continuity of operations plan assessments that independently evaluates executive branch agencies' compliance with Federal Preparedness Circular (FPC) 65, including the extent to which agencies (1) identify essential functions; (2) identify the levels of staff and resources required at their alternate facilities; and (3) plan, conduct, and document the necessary tests and exercises at the appropriate scope and frequency. The methodology should include a mechanism for reporting any deficiencies to the head of the agency for correction. • Establish a timeline for developing, in consultation with the Office of Personnel Management (OPM), guidance on the steps that agencies should take to adequately prepare for the use of telework during a continuity of operations event.	DHS partially agreed and stated that FEMA would be conducting assessments of the six agencies in conjunction with its upcoming interagency exercise. DHS stated that FEMA currently has an assessment program and a methodology that includes (1) a self-assessment tool to assist senior leaders in performing internal assessments, (2) the government wide exercise planned for June 2006, and (3) a plan for an expanded comprehensive department and agency assessment program. DHS stated that the agency's efforts are constrained by resources, and that additional resources are required to support the large number of federal offices and facilities that could benefit from recurring continuity of operations planning assessments. While these steps outline an overall approach to assessing agency plans, they do not constitute a methodology for assessing agency compliance with FPC 65. Without such a methodology, the agency will have limited assurance that agencies have taken the necessary steps to prepare for an emergency. In responding to our recommendation on developing guidance on agencies' use of telework, DHS partially agreed and stated that FEMA will coordinate with OPM in the development of a timeline for further telework guidance. However, as stated in our report, present guidance does not address the preparations agencies should make for using telework during emergencies and it is unclear whether and when FEMA will release such guidance.

Table 5. (Continued).

Title and GAO product number	Summary of recommendations	Agency response
Hurricanes Katrina and Rita: Coordination between FEMA and the Red Cross Should Be Improved for the 2006 Hurricane Season, GAO-06-712	Recommendations to the Secretary of Homeland Security: To clarify roles and responsibilities within ESF-6 for the 2006 hurricane season and to help ensure that FEMA's resource tracking system will meet the needs of those requesting FEMA assistance, the Secretary of DHS should • Direct FEMA to work with the Interim President and Chief Executive Officer of the Red Cross as soon as possible to reach agreement on the operating procedures that they will both use in the event of an incident of national significance. Given the lack of progress FEMA and the Red Cross have made thus far in reaching agreement on the operating procedures and that the new hurricane season is beginning, they may wish to use mediation to speed the agreement. Direct FEMA to ensure that it obtains input from the Red Cross as it develops a resource tracking system. • Recommendations to the Interim President and Chief Executive Officer of the Red Cross: • Implement ESF-6 staffing strategies that better facilitate the development of working relationships and retain institutional knowledge.	DHS did not provide a response to our recommendations, noting that FEMA was actively preparing for the hurricane season. Overall, the Red Cross agreed with our conclusion that coordination between FEMA and the Red Cross could be improved for the 2006 hurricane season. The Red Cross also highlighted actions under way with respect to our first two recommendations. With respect to our recommendation about staffing strategies, the Red Cross said that it is in the process of hiring ESF-6 reservists who will be deployed for extended periods of time to perform Red Cross ESF-6 mass care functions at the federal level.
Expedited Assistance for Victims of Hurricanes Katrina and Rita: FEMA's Control Weaknesses Exposed the Government to Significant Fraud and Abuse, GAO-06-655	Recommendations to the Secretary of Homeland Security: That the Secretary of Homeland Security directs Undersecretary for Federal Emergency Management to take six actions to address the weaknesses we identified in the administration of Individuals and Households Program (IHP). • Establish an identity verification process for IHP registrants applying via both the Internet and telephone, to provide reasonable assurance that disaster assistance payments are made only to qualified individuals. Within this process • establish detailed criteria for registration and provide clear instructions to registrants on the identification information required, • create a field within the registration that asks registrants to provide their name exactly as it appears on their social security card in order to prevent name and social security number (SSN) mismatches, • fully field test the identity verification process prior to implementation, • ensure that call center employees give real-time feedback to registrants on whether their identities have been validated, and • establish a process that uses alternative means of identity verification to expeditiously handle legitimate applicants that are rejected by identity verification controls. • Develop procedures to improve the existing review process of duplicate registrations containing the exact same SSN and to identify the reasons why registrations flagged	DHS and FEMA concurred fully with four of our six recommendations, and partially concurred with the remaining two recommendations. FEMA and DHS concurred fully that FEMA (1) improve procedures to review registrations containing the same SSNs and other duplicate information; (2) subject all registration addresses to verification during the registration process; (3) explore entering into agreements with other agencies, such as the Social Security Administration, to periodically authenticate IHP information; and (4) issue proper instructions to any future debit card recipients. FEMA and DHS stated that they have already taken actions to address these recommendations. These actions include instituting an Internet application process that will prevent all duplicate registrations from the Internet, implementing procedures so that call centers will no longer accept duplicate registrations with the same SSN in the same disaster, and conducting conference calls and conducting data sharing tests with the Social

Table 5. (Continued).

Title and GAO product number	Summary of recommendations	Agency response
	as invalid or as potential duplicates have been overridden and approved for payment. • Establish an address verification process for IHP registrants applying via both the Internet and telephone, to provide reasonable assurance that disaster assistance payments are made only to qualified individuals. Within this process • create a uniform method to input street names and numbers and apartment numbers into the registration, • institute procedures to check IHP registration damaged addresses against publicly available address databases so that payments are not made based on bogus property addresses, • fully field test the address verification process prior to implementation, • ensure that call center employees can give real time feedback to registrants on whether addresses have been validated, and • establish a process that uses alternative means of address verification to expeditiously handle legitimate applicants that are rejected by address verification controls. • Explore entering into an agreement with other agencies, such as the Social Security Administration, to periodically authenticate information contained in IHP registrations. • Establish procedures to collect duplicate expedited assistance payments or to offset these amounts against future payments. Such duplicate payments include • the payments made to IHP recipients who improperly received the $2,000 debit cards and an additional $2,000 Expedited Assistance check or Electronic Funds Transfer, and • the thousands of duplicate Expedited Assistance payments made to the same IHP registration number. • Ensure that any future distribution of IHP debit cards includes instructions on the proper use of IHP funds, similar to those instructions provided to IHP check and Electronic Funds Transfer recipients, to prevent improper usage.	Security Administration. In addition, DHS and FEMA stated that, starting in June 2006, all registration addresses (even phone-in) will be subjected to an online verification during the registration process. While these are steps in the right direction, we will follow up on whether the actions taken fully address our recommendations. FEMA and DHS partially concurred with our recommendation concerning duplicate payments. FEMA and DHS took exception with our categorization of some payments as being potential duplicates, and with our assessment that they should initiate actions to collect duplicate Expedited Assistance payments. However, for all our case study examples, we conducted further investigative work to confirm that the payments were made to actual duplicates, not covered by the separated household policy, and were therefore improper payments. As for initiating actions to collect duplicate payments, DHS and FEMA stated that they had processed for recoupment nearly all the payments they believed were duplicates as of April 1, 2006. While we have not assessed the effectiveness of FEMA's recoupment process, we continue to believe that FEMA should attempt to recoup as many dollars of improper payments as possible, including those duplicate payments that we identified that FEMA questioned. Going forward it will be important for FEMA to establish effective controls to prevent fraudulent and improper payments before they occur, because fraud prevention is a far more effective control than detecting improper and potentially fraudulent

Table 5. (Continued).

Title and GAO product number	Summary of recommendations	Agency response
		payments after they are made. Our experience with organizations that rely on a process that attempts to detect improper and potentially fraudulent payments after they are made is that the organization recovers only a fraction of the payments that should not have been made.
Hurricane Katrina: Better Plans and Exercises Needed to Guide the Military's Response to Catastrophic Natural Disasters, GAO-06-643	Recommendations to the Secretary of Defense: To improve the military response to catastrophic disasters, the Secretary of Defense should • Provide proposed revisions of the NRP to the Secretary of the Department of Homeland Security that addresses the proactive functions the military is expected to make during a catastrophic incident. • Establish milestones and expedite the development of detailed plans and exercises to fully account for the unique capabilities and support that the military is likely to provide to civil authorities in response to the full range of domestic disasters, including catastrophes. The plans and exercises should specifically address the • use of reconnaissance capabilities to assess damage, • use of communications capabilities to facilitate support to civil authorities, • integration of active component and National Guard and Reserve forces, • use of search and rescue capabilities and the military's role in search and rescue, and • role the military might be expected to play in logistics. • Direct the Chief of the National Guard Bureau to work with the state governors and adjutants general to develop and maintain a list of the types of capabilities the National Guard will likely provide in response to domestic natural disasters under state-to-state mutual assistance agreements along with the associated units that could provide these capabilities. This information should be made available to the Northern Command, U.S. Joint Forces Command, and other organizations with federal military support to civil authority planning responsibilities. • Establish milestones and identify the types of scalable federal military capabilities and the units that could provide those capabilities in response to the full range of domestic disasters and catastrophes covered by DOD's defense support to civil authorities' plans.	DOD commented on our four recommendations, partially concurring with each of them. With respect to our first recommendation—to revise the NRP to fully address the proactive functions that the military will be expected to perform during a catastrophic incident—DOD said that proactive military functions can be identified in all 15 major disaster scenarios and said it is working with the Department of Homeland Security to revise the NRP. While DOD stated that the long-term focus of the U.S. government should be to develop more robust domestic disaster capabilities within the Department of Homeland Security, it acknowledged that DOD will need to assume a more robust response role in the interim period and when other responders lack the resources and expertise to handle a particular disaster. With respect to our second recommendation that concerned the development of detailed plans and exercises, DOD listed a number of steps it is taking to improve its disaster response planning and exercises and said that consistent with its Strategy for Homeland Defense and Civil Support the active component should complement, but not duplicate, the National Guard's likely role as an early responder. On June 27, 2006, the Assistant Secretary of Defense for Homeland Defense sent DOD's final response, listing some recently-completed exercises as well as steps that DOD was taking to enhance interagency planning efforts.

Table 5. (Continued).

Title and GAO product number	Summary of recommendations	Agency response
	Matters for Congressional Consideration: In a 1993 report we suggested that Congress consider removing the statutory restriction on DOD's authority to involuntarily activate Reserve units for catastrophic disaster relief. In view of the significant military downsizing that has occurred since we first raised this matter and the need to actively engage the total force in order to meet missions at home and abroad, we continue to believe that Congress should consider lifting or modifying the mobilization restriction—10 U.S.C. §12304 (c)(1)—that limits reserve component participation in catastrophic natural disasters.	DOD also partially concurred with our third recommendation—that the Chief of the National Guard Bureau work with the state governors and adjutants general to develop and maintain a list of the types of capabilities the National Guard will likely provide in response to domestic natural disasters under state-to-state mutual assistance agreements. Finally, DOD partially concurred with our recommendation that it identify the types of scalable federal military capabilities and units that will provide those capabilities in response to the full range of domestic disasters and catastrophes. In its final response to our report, DOD stated it had analyzed potential gaps between state and other federal agency response, and had developed 18 pre-scripted mission assignments to address DOD support to FEMA, and had validated these assignments during an exercise. DOD also reported that it had developed scalable capability packages in conjunctions with U.S. Northern Command's contingency plan for defense support to civil authorities, and it was staffing forces to accelerate force/capabilities necessary to support a catastrophic event.

Table 5. (Continued).

Title and GAO product number	Summary of recommendations	Agency response
U.S. Tsunami Preparedness: Federal and State Partners Collaborate to Help Communities Reduce Potential Impacts, but Significant Challenges Remain, GAO-06-519	Recommendations to the Secretary of Commerce: To help improve national tsunami preparedness, the Secretary of Commerce should direct the National Oceanographic and Atmospheric Administration (NOAA) Administrator to take the following six actions • Work with the FEMA Director and the United States Geological Survey (USGS) Director to create standardized tsunami loss estimation software to help communities determine the potential impact of tsunamis and identify appropriate mitigation actions. • Reduce the number of tsunami warning false alarms by (1) completing the planned expansion of tsunami detection stations; (2) reexamining NWS rules dictating when a warning will be issued and to which areas; (3) establishing a routine process for other federal and state experts to formally review and comment on the centers' use of seismic data; and (4) setting performance goals to guide improvements. • Work with the states to conduct periodic end-to-end tests of the tsunami warning system, including NOAA Weather Radio and the Emergency Alert System, to ensure the system will function as intended during a tsunami emergency. • Evaluate the Tsunami Ready program to determine what barriers, if any, exist to participation and what modifications are needed to encourage more high-risk communities to participate. • Evaluate the National Tsunami Hazard Mitigation Program to determine what has worked well in the past and what high priority activities remain to be completed and to help inform strategic planning efforts. • Develop comprehensive risk-based strategic plans for the Tsunami Program and National Tsunami Hazard Mitigation Program that consider input from states and federal partners and include metrics for measuring progress toward achieving program goals.	Commerce, representing NOAA, concurred with all six recommendations. However, in NOAA suggested a revision to one of the recommendations with which we disagree. In response to our recommendation that NOAA evaluate the Tsunami Ready program to determine what barriers, if any, exist to participation and what modifications are needed to encourage more high-risk communities to participate, NOAA suggested changing the recommendation's focus from "high-risk" to "at-risk" communities. According to NOAA all U.S. coastal communities should be prepared for a tsunami no matter how rare. While we agree that preparing all U.S. coastal communities for a tsunami may be a laudable long-term goal, given the agency's limited resources, it may be an unrealistic goal in the short-term. Therefore, we believe that NOAA should use a risk-based approach and target initial participation in the Tsunami-Ready program to those communities that face the greatest risk. Homeland Security, representing FEMA, commented on one of the six recommendations and indicated that while it concurred with the recommendation, that NOAA work with FEMA and USGS to create standardized tsunami loss estimation software, it was concerned that FEMA did not have the funding or the staff resources to pursue such a request and that such a request from NOAA would have to address these resource needs. The Department of the Interior commented that the report was a thorough, well-researched examination of the nation's tsunami warning system and that it correctly recognizes the need for close collaboration at the federal, state, and local levels to have an effective tsunami warning system. Interior also said that it supports the need for a risk-based approach to prioritizing federal investments in this system and is actively collaborating with NOAA to provide the

Table 5. (Continued).

Title and GAO product number	Summary of recommendations	Agency response
		hazard assessments necessary for such an approach.
Hurricane Katrina: Comprehensive Policies and Procedures Are Needed to Ensure Appropriate Use of and Accountability for International Assistance, GAO-06-460	Recommendations to the Secretary of Homeland Security: To help ensure that the cognizant agencies fulfill their responsibility to account for and effectively manage foreign donations and maintain adequate internal controls over government resources, the Secretary of Homeland Security, in consultation with the Secretary, Department of State, should: • Establish within the NRP—or other appropriate plans—clearly delineated policies and procedures for the acceptance, receipt, and distribution of international assistance. • Incorporate the following actions and procedures into their guidance: • Develop policies, procedures, and plans to help ensure international cash donations for disaster relief and assistance are accepted and used appropriately as needed, • Place international cash donations in an account that would pay interest while decisions are pending on their use to maintain the purchasing power of those donations, • Maintain oversight of foreign donated in-kind assets by tracking them from receipt to disbursement, to reasonably ensure that assistance is delivered where it is intended, and • Establish plans for the acceptance of foreign-donated items that include coordinating with regulatory agencies, such as US Department of Agriculture and Food and Drug Administration, in advance, in order to prevent the acceptance of items that are prohibited from distribution in the United States regardless of waivers that might be established to expedite the importing of foreign assistance; these plans should also include Department of State obtaining information on acceptable or unacceptable items in order to communicate to the international community what is needed or what can not be accepted. Recommendations to the Secretary of Defense: The Secretary of Defense, in consultation with the Secretaries of State and Homeland Security, should: • Establish within the NRP—or other appropriate plans—clearly delineated policies and procedures to ensure that foreign military offers of assistance for domestic disasters are coordinated through the Department of State to ensure they are properly accepted and safeguarded and used as intended. Develop and issue internal DOD guidance to commanders on the agreed-upon process to coordinate assistance through Department of State.	DOD and DHS generally agreed with our recommendations. DHS noted that, in some cases, actions were already underway to address the recommendations. Both made suggestions to clarify the wording, and we adjusted the recommendations based on their suggestions.

Table 5. (Continued).

Title and GAO product number	Summary of recommendations	Agency response
Federal Emergency Management Agency: Improvements Needed to Enhance Oversight and Management of the National Flood Insurance Program, GAO-06-119; Future Financial Stability of the National Flood Insurance Program, GAO-06-174T; and Oversight and Management of the National Flood Insurance Program, GAO-06-183T	Recommendations to the Secretary of Homeland Security: To improve FEMA's oversight and management of the NFIP and make it more actuarially sound, we recommend that the Secretary of the Department of Homeland Security direct the Undersecretary for Emergency Preparedness and Response to: • Use a methodologically valid approach to draw statistically representative samples of claims for underwriting and claims portions of operational reviews and for quality assurance re-inspections of claims by general adjusters. • Develop documented plans with milestones for implementing requirements of the Flood Insurance Reform Act of 2004 to provide policyholders with a flood insurance claims handbook that meets statutory requirements, to establish a regulatory appeals process, and to ensure that insurance agents meet minimum NFIP education and training requirements.	FEMA offered comments principally in three areas: (1) its disappointment that we had not directly addressed the issue of whether Congress intended the flood insurance program to restore damaged property to its pre-flood condition; (2) its view that the method of choosing its sample for operational reviews was appropriate and that its financial and internal controls are wide-ranging and include processes that we did not address; and (3) its view that contrary to the impression given in our draft report, FEMA has worked diligently to implement the requirements of the Flood Insurance Reform Act of 2004. We responded that (1) we believed that we have addressed the issue of congressional intent consistent with our statutory mandate by explaining the statutory and regulatory provisions that affect both dollar ceilings and other coverage limitations; (2) FEMA's sampling and internal controls still did not provide management with the information needed to assess the overall performance of companies, the overall accuracy of the underwriting of NFIP policies and the adjustment of claims—information that FEMA needs to have reasonable assurance that program objectives are being achieved; and (3) we described several actions FEMA had taken in its efforts to comply with the act, while noting that it had not fully implemented the act's requirements.

Table 6. New GAO Recommendations Formalized in this Report

We are making several additional recommendations regarding preparedness, response, and recovery: Recommendations to the Secretary of Homeland Security: The Secretary should: • Rigorously re-test, train, and exercise its recent clarification of the roles, responsibilities, and lines of authority for all levels of leadership, implementing changes needed to remedy identified coordination problems. • Direct that the NRP base plan and its Catastrophic Incident Annex be supported by more robust and detailed operational implementation plans, particularly the Catastrophic Incident Supplement to the NRP. Such operational plans should, for example, further define and leverage those military capabilities that might be needed in a catastrophic disaster. • Provide guidance and direction for federal, state, and local planning, training, and exercises to ensure such activities fully support preparedness, response, and recovery responsibilities at a jurisdictional and regional basis. This should also include the application of lessons learned from actual catastrophic and other disasters. • Take the lead in monitoring federal agencies' efforts to meet their responsibilities under the NRP and the interim National Preparedness Goal, including the development, testing, and exercising of agency operational plans to implement their responsibilities under the NRP, NIMS, and the National Preparedness Goal. • Given that resources are finite, apply an all-hazards risk management approach in deciding whether and how to invest in specific capabilities for a catastrophic disaster. • Provide guidance on advance procurement practices and procedures for those federal agencies with roles and responsibilities under the NRP, so that these agencies can better manage disaster-related procurements. These practices should be in advance of disasters, ongoing and continuous, and include (1) developing knowledge of contractor capabilities, and available commodities, services and prices as well as developing pre-established vendor relationships, on a competitive basis whenever feasible; (2) establishing scalable operations plans to adjust the level of capacity needed to respond; (3) formally assigning and communicating disaster-related responsibilities and, where feasible, incorporating necessary training; and (4) providing sufficient numbers of field-level contracting staff to meet mission requirements. DHS should also establish an assessment process to monitor agencies' continuous planning efforts for their disaster related procurement needs and the maintenance of capabilities. Matter for Congressional Consideration: • We again recommend, as we did in 1993 in the aftermath of Hurricane Andrew, that Congress give federal agencies explicit authority to take actions to prepare for catastrophic disasters when there is warning.

Source: GAO analysis.

REFERENCES

[1] A Failure of Initiative: Final Report of the House Select Bipartisan Committee to Investigate the Preparation for And Response to Hurricane Katrina (Washington, D.C.: Feb. 15, 2006).

[2] Hurricane Katrina: A Nation Still Unprepared (Washington, D.C.: May 2006).

[3] The Federal Response to Hurricane Katrina: Lessons Learned (Washington, D.C.: Feb. 23, 2006).

[4] A Performance Review of FEMA's Disaster Management Activities in Response to Hurricane Katrina, OIG-06-32 (Washington, D.C.: Mar. 31, 2006).

[5] DHS/FEMA Initial Response Hotwash: Hurricane Katrina in Louisiana, DR-1603-LA (Baton Rouge, La.: Feb. 13, 2006).

[6] GAO, Hurricane Katrina: GAO's Preliminary Observations Regarding Preparedness, Response, and Recovery, GAO-06-442T (Washington, D.C.: Mar. 8, 2006).

[7] GAO, Disaster Management: Improving the Nation's Response to Catastrophic Disasters, GAO/RCED-93-186 (Washington, D.C.: July 23, 1993).

[8] GAO, Disaster Assistance: Information on FEMA's Post 9/11 Public Assistance to the New York City Area, GAO-03-926 (Washington, D.C.: Aug. 29, 2003); and GAO, September 11: Overview of Federal Disaster Assistance to the New York City Area, GAO-04-72 (Washington, D.C.: Oct. 31, 2003).

[9] GAO, U.S. Tsunami Preparedness: Federal and State Partners Collaborate to Help Communities Reduce Potential Impacts, but Significant Challenges Remain, GAO-06-519 (Washington, D.C.: June 5, 2006); and Foreign Assistance: USAID Has Begun Tsunami Reconstruction in Indonesia and Sri Lanka, but Key Projects May Exceed Initial Cost and Schedule Estimates, GAO-06-488 (Washington, D.C.: Apr. 14, 2006).

[10] GAO, Hurricane Katrina: Better Plans and Exercises Needed to Guide the Military's Response to Catastrophic Natural Disasters, GAO-06-643 (Washington, D.C.: May 15, 2006).

[11] See GAO, Agency Management of Contractors Responding to Hurricanes Katrina and Rita, GAO-06-461R (Washington, D.C.: Mar. 16, 2006); Hurricane Katrina: Planning for and Management of Federal Disaster Recovery Contracts, GAO-06-622T (Washington, D.C.: Apr. 10, 2006); and Hurricane Katrina: Improving Federal Contracting Practices in Disaster Recovery Operations, GAO-06-714T (Washington, D.C.: May 4, 2006).

[12] See GAO, Disaster Preparedness: Preliminary Observations on the Evacuation of Hospitals and Nursing Homes Due to Hurricanes, GAO-06-443R (Washington, D.C.: Feb. 16, 2006); and Disaster Preparedness: Preliminary Observations on the Evacuation of Vulnerable Populations due to Hurricanes and Other Disasters, GAO-790T (Washington, D.C.: May 18, 2006); Disaster Preparedness: Limitations in Federal Evacuation Assistance for Health Facilities Should be Addressed, GAO-06-826 (Washington, D.C.: July 20, 2006).

[13] See GAO, Hurricanes Katrina and Rita: Provision of Charitable Assistance, GAO-06-297T (Washington, D.C.: Dec. 13, 2005); and Hurricanes Katrina and Rita : Coordination between FEMA and Red Cross Should Be Improved for the 2006 Hurricane Season, GAO-06-712 (Washington, D.C.: June 8, 2006).

[14] GAO, Hurricane Katrina: Comprehensive Procedures Are Needed to Ensure Appropriate Use of and Accountability for International Assistance, GAO-06-460 (Washington, D.C.: Apr. 6, 2006).

[15] GAO, Expedited Assistance for Victims of Hurricanes Katrina and Rita: FEMA's Control Weaknesses Exposed the Government to Significant Fraud and Abuse, GAO-06-655 (Washington, D.C.: June 16, 2006).

[16] The National Response Plan is the basis for how federal departments and agencies are to work together with state, local, and tribal governments and the private sector in managing domestic incidents.

[17] 42 U.S.C. §§ 5121-5206.

[18] Pub. L. No. 104-321, 110 Stat. 3877 (1996).

[19] The National Emergency Management Association is a non-partisan, non-profit 501(c)(3) association affiliated with The Council of State Governments.

[20] Pub. L. No. 107-296, 116 Stat. 2135 (2002).

[21] Establishment of a Coordinator of Federal Support for the Recovery and Rebuilding of the Gulf Coast Region," Exec. Order No. 13390, 70 Fed. Reg. 67327 (2005).

[22] The NRP revisions include a Quick Reference Guide that provides a condensed, 23-page overview of the NRP as modified.

[23] The HSPD-5 criteria are: (1) a federal department or agency acting under its own authority has requested the assistance of the Secretary of Homeland Security; (2) the resources of state and local authorities are overwhelmed and federal assistance has been requested by the appropriate state and local authorities, for example, under the Stafford Act; (3) more than one federal department or agency has become substantially involved in responding to an incident; or (4) the Secretary of Homeland Security has been directed by the President to assume responsibility for managing a domestic incident.

[24] According to the Catastrophic Incident Annex, all catastrophic incidents are Incidents of National Significance. Implementation of the annex requires a separate catastrophic incident designation by the Secretary of Homeland Security.

[25] See GAO, Disaster Management: Improving the Nation's Response to Catastrophic Disasters, GAO/RCED-93-186 (Washington, D.C.: July 23, 1993); and Disaster Management: Recent Disasters Demonstrate the Need to Improve the Nation's Response Strategy, GAO/T-RCED-93-46 (Washington, D.C.: May 25, 1993).

[26] GAO, Hurricane Katrina: GAO's Preliminary Observations Regarding Preparedness, Response, and Recovery, GAO-06-442T (Washington, D.C.: Mar. 8, 2006).

[27] The National Search and Rescue Plan (1999) provides an overall plan for coordination of search and rescue operations, the effective use of all available resources, mutual assistance, and efforts to improve such cooperation and services. The goal of the plan is to integrate available resources that can be used for search and rescue into a cooperative network for greater protection of life and property and to ensure greater efficiency and economy.

[28] GAO-06-643.

[29] GAO, Hurricanes Katrina and Rita: Coordination between FEMA and the Red Cross Should Be Improved for the 2006 Hurricane Season, GAO-06-712 (Washington, D.C.: June 8, 2006).

[30] GAO, Homeland Security: Critical Design and Implementation Issues GAO-02-957T (Washington, D.C.: July 17, 2002).

[31] As a result of the Secretary's Second Stage Review, the Director of FEMA also became the Undersecretary of Emergency Management.

[32] See United States Emergency Management Authority Act of 2006, S. 3595, 109th Cong. § 504 (2006); National Emergency Management Reform and Enhancement Act of 2006, H.R. 5351, 109th Cong. § 502(b) (2006).

[33] GAO, Homeland Security: DHS' Efforts to Enhance First Responders' All-Hazards Capabilities Continue to Evolve, GAO-05-652 (Washington, D.C.: July 11, 2005).

[34] See GAO, Disaster Management: Improving the Nation's Response to Catastrophic Disasters, GAO-93-186 (Washington, D.C.: July 23, 1993); and Disaster Management: Recent Disasters Demonstrate the Need to Improve the Nation's Response Strategy, GAO-93-46 (Washington, D.C.: May 25, 1993).

[35] GAO-06-643.

[36] GAO-06-643.

[37] Typically, military equipment cannot communicate with civilian police, fire, and emergency medical systems unless it is augmented with specialized equipment. During Hurricane Katrina, the military and civilian agencies deployed mobile communication vans that were able to connect different communications systems that are normally incompatible.

[38] GAO, Homeland Security: Federal Leadership and Intergovernmental Cooperation Required to Achieve First Responder Interoperable Communications, GAO-04-740 (Washington, D.C.: July 20, 2004).

[39] Federal Communications Commission, Notice of Proposed Rulemaking, FCC-06-83 (Washington, D.C.: June 19, 2006).

[40] The September 2005 draft Catastrophic Incident Supplement to the NRP is intended to be used with the Catastrophic Incident Annex when a catastrophic incident almost immediately overwhelms the capabilities of state and local governments. The draft supplement does not describe what, if any, role the federal government is to play in coordinating with state and local authorities for this kind of transportation.

[41] GAO, Disaster Evacuations: Limitation in Federal Assistance to Health Facilities for Transportation Should Be Addressed, GAO-06-826 (Washington, D.C.: July 20, 2006).

[42] GAO, Lessons Learned for Protecting and Educating Children after the Gulf Coast Hurricanes, GAO-06-680R (Washington, D.C.: May 11, 2006).

[43] We observed that while the reasons children were separated from their families are not fully understood, in many cases children were evacuated separately from parents and

sent to different shelters. All of the instances of children reported missing to the National Center for Missing and Exploited Children were resolved by March 2006. All of the children reported missing were not necessarily unaccompanied or in harm's way; some who were safe were probably reported as missing because non-guardians (such as neighbors) could report a child as missing.

[44] GAO-06-643.

[45] Pub. L. No. 109-59, 119 Stat. 1144, 1934 (2005).

[46] H.R. Rep. No. 109-241, at 68 (2006).

[47] As used in the Nationwide Review Plan, the term "state" refers to any state of the United States, the District of Columbia, the Commonwealth of Puerto Rico, the Virgin Islands, Guam, American Samoa, the Commonwealth of the Northern Mariana Islands, and any possession of the United States. The term "urban areas" refers to the 55 Fiscal Year 2005 Urban Areas Security Initiative program grantees; and the 20 major cities selected for the Nationwide Plan Review by DHS were based on a DHS analysis of 2004 population, risk, and need.

[48] GAO-06-643.

[49] GAO-06-403T.

[50] GAO, Small Business Administration: Actions Needed to Provide More Timely Disaster Assistance, GAO-06-860 (Washington, D.C.: July 28, 2006).

[51] We surveyed 50 states, the District of Columbia, and Puerto Rico; 20 states and the District of Columbia reported that they had a written plan, 17 states and Puerto Rico reported they did not have a written plan, and 11 states did not respond to the disaster planning questions in our survey.

[52] GAO, Federal Action Needed to Ensure States Have Plans to Safeguard Children in the Child Welfare System Displaced by Disasters. GAO-06-944 (Washington, DC.: July 28, 2006).

[53] See GAO, Disaster Assistance: DOD's Support for Hurricanes Andrew and Iniki and Typhoon Omar, GAO/NSIAD-93-180 (Washington, D.C.: June 18, 1993); and Disaster Assistance: Federal, State, and Local Responses to Natural Disasters Need Improvement, GAO/RCED-91-43 (Washington, D.C.: Mar. 6, 1991).

[54] GAO, Continuity of Operations: Selected Agencies Could Improve Planning for Use of Alternate Facilities and Telework during Disruptions, GAO-06-713 (Washington, D.C.: May 11, 2006).

[55] GAO-06-643.

[56] GAO-05-652.

[57] GAO, Homeland Security: DHS's Efforts to Enhance First Responders' All-Hazards Capabilities Continue to Evolve, GAO-05-652 (Washington, D.C.: July 11, 2005).

[58] GAO, Statement by Comptroller General David M. Walker on GAO's Preliminary Observations Regarding Preparedness and Response to Hurricanes Katrina and Rita, GAO-06-365R. (Washington: D.C.: Feb. 1, 2006).

[59] Four emergency supplemental appropriations acts account for the approximately $88 billion in hurricane relief. See Emergency Supplemental Appropriations Act to Meet Immediate Needs Arising From the Consequences of Hurricane Katrina, 2005, Pub. L. No. 109-61, 119 Stat. 1988 (2005); Second Emergency Supplemental Appropriations Act to Meet Immediate Needs Arising From the Consequences of Hurricane Katrina, 2005, Pub. L. No. 109-62, 119 Stat. 1990 (2005); Emergency Supplemental

Appropriations to Address Hurricanes in the Gulf of Mexico, Pub. L. No. 109-148, div. B, title I, 119 Stat. 2680, 2745 (2005); and Further Hurricane Disaster Relief and Recovery, Pub. L. No. 109-234, title II, 120 Stat. 418, 443 (2006).
[60] One contractor provided ice production and transport services, another provided cold storage and transport services, and a third contractor provided a variety of transportation services, including refrigerated trucks.
[61] GAO, Hurricane Katrina: Army Corps of Engineers Contract for Mississippi Classrooms, GAO-06-454 (Washington D.C.: May 1, 2006).
[62] GAO, Coast Guard: Observations on the Preparation, Response, and Recovery Missions Related to Hurricane Katrina, GAO-06-903 (Washington D.C.: July 31, 2006).
[63] 42 U.S.C. § 5201(b).
[64] GAO, Hurricane Katrina: Comprehensive Policies and Procedures Are Needed to Ensure Appropriate Use of and Accountability for International Assistance, GAO-06-460 (Washington D.C.: Apr. 6, 2006).
[65] FEMA may extend this 18-month limit if it determines that, due to extraordinary circumstances, an extension would be in the public interest. 44 C.F.R. § 206.110(e).
[66] The maximum of IHP assistance is statutorily capped at $25,000, adjusted annually to reflect changes in the Consumer Price Index. 42 U.S.C § 5174(h). In 2005, the maximum was $26,200.
[67] Loans from the Small Business Administration are considered to be the primary means of disaster assistance for disaster victims who have the financial ability to repay such loans.
[68] GAO, Expedited Assistance for Victims of Hurricanes Katrina and Rita: FEMA's Control Weaknesses Exposed the Government to Significant Fraud and Abuse, GAO-06-403T (Washington, D.C.: Feb. 13, 2006).
[69] GAO, Hurricanes Katrina and Rita: Improper and Potentially Fraudulent Individual Assistance Payments Estimated to Be between $600 Million and $1.4 Billion, GAO-06-844T (Washington, D.C.: June 14, 2006).
[70] Expedited Assistance—a component of the IHP program during Hurricanes Katrina and Rita—took the form of $2,000 payments to disaster victims to help with the immediate, emergency needs for food, shelter, clothing, and personal necessities.
[71] GAO-06-844T.
[72] GAO, Expedited Assistance for Victims of Hurricanes Katrina and Rita: FEMA's Control Weaknesses Exposed the Government to Significant Fraud and Abuse, GAO-06-655 (Washington, D.C.: June 16, 2006).
[73] Pub. L. No. 109-148.
[74] Pub. L. No. 109-234.
[75] Section 408 of the Stafford Act.
[76] Eligibility requirements for receiving assistance under IHP include demonstrating that an applicant's pre-disaster residence is located in the declared disaster area; that damage to the property was not covered by insurance; and that the applicant is unable to return to their pre-disaster home due to damage caused by the disaster.
[77] FEMA announced certain exceptions to the May 31 deadline. In particular, FEMA stated that it planned to make every effort to notify states of the IHP eligibility status of evacuees before April 15. If this notification occurred after April 15, states could

request additional time to provide eligible and ineligible evacuees with a 30-day lease termination notice. Specifically, states could receive up to 15 additional days for eligible evacuees, and up to 30 additional days for ineligible evacuees. Also, for ineligible evacuees, the state would be reimbursed for contractual lease termination costs associated with leases that require greater than 30 days notice. FEMA subsequently extended the May 31 deadline to June 30 for 11 jurisdictions. FEMA further extended the deadline for one of these jurisdictions—Houston, Texas—until September 30th.

[78] The Department of Defense Appropriations Act, 2006 (P.L. 109-148).

[79] Under the Disaster Voucher Program, funding must be obligated by September 30, 2007. Unless expressly renewed or extended by law, assistance may not continue beyond the time the obligated funds are expended.

[80] The National Flood Insurance Act of 1968. The act, as amended, is codified at 42 U.S.C. §§ 4001 to 4129.

[81] NFIP coverage is also available for other structures such as apartment buildings, schools, churches, businesses and condominium associations, but the coverage terms differ in various respects from homeowners' coverage.

[82] GAO, Federal Emergency Management Agency: Challenges for the National Flood Insurance Program, GAO-06-335T (Washington, D.C.: Jan. 25, 2006).

[83] The NFIP standard flood insurance policy considers a property substantially damaged if the cost of repairing it exceeds 50 percent of its market value at the time of the flood.

[84] GAO, Flood Map Modernization: Program Strategy Shows Promise, but Challenges Remain, GAO-04-417 (Washington, D.C.: Mar. 31, 2004).

[85] GAO, Federal Emergency Management Agency: Improvements Needed to Enhance Oversight and Management of the National Flood Insurance Program, GAO-06-119 (Washington, D.C.: Oct. 18, 2005).

[86] In response to the Flood Insurance Reform Act of 2004, FEMA recently published an interim final rule establishing a regulatory appeals process for flood insurance claimants. National Flood Insurance Program; Appeal of Decisions Relating to Flood Insurance Claims, 71 Fed. Reg. 30294 (2006).

[87] GAO, Guidelines for Rescuing Large Failing Firms and Municipalities, GAO/GGD-84-34 (Washington, D.C.: Mar. 29, 1984).

[88] GAO, Homeland Security: DHS' Effort to Enhance First Responders' All-Hazards Capabilities Continue to Evolve, GAO-04-652 (Washington, D.C.: July 11, 2005).

[89] GAO, Limitations in Federal Evacuation Assistance for Health Facilities Should be Addressed, GAO-06-826 (Washington, D.C.: July 20, 2006).

[90] GAO, Small Business Administration: Actions Needed to Provide More Timely Assistance, GAO-06-860 (Washington, D.C.: July 28, 2006)

INDEX

A

B

C

G

H

I

T

U

V

W